DATE DUE

MAR 2 1 1995 Ap 3	
JAN 2 3 1996	
Mar 4	
MAR 2 9 1996	
OCT 2 3 1996	
NOV - 8 1996	
NOV 2 6 1996	
FEB 1 2 1997	
FEB 2 1 1997	
Mar 7	
DEC - 3 1998	
MAR 1 5 1999	
OCT 2 3 2001	

BRODART.   Cat. No. 23-221

NEUROPSYCHOLOGICAL DIFFERENTIATION
OF DEMENTIA SYNDROMES

Neuropsychological Differentiation of Dementia Syndromes

Mayke M.A. Derix, Ph.D.

Department of Psychiatry and Neuropsychology
Section of Neuropsychology, Neuropsychiatry and Psychobiology
Limburg University
P.O. Box 616
6200 MP Maastricht
The Netherlands

SWETS & ZEITLINGER

LISSE ❙ BERWYN, PA   ❙ ACADEMIC PUBLISHING DIVISION ❙

Library of Congress Cataloging-in-Publication Data

(applied for)

Cip-gegevens Koninklijke Bibliotheek, Den Haag

Derix, Mayke M.A.

Neuropsychological differentiation of dementia syndromes /
Mayke, M.A. Derix. - Lisse [etc.] : Swets & Zeitlinger. -
Ill.
Ook verschenen als proefschrift Universiteit van Amsterdam, 1991. -
Met index, lit. opg.
ISBN 90-265-1312-7
NUGI 742
Trefw.: dementie ; neuropsychologie.

Cover design: Rob Molthoff, Lin*ea*Forma, Alkmaar
Cover printed in the Netherlands by Casparie, IJsselstein
Printed in the Netherlands by
Offsetdrukkerij Kanters B.V., Alblasserdam

ISBN 90-265-1312-7
NUGI 742

Foreword

This timely book began as a doctoral dissertation in neuro-psychology but it differs from many that fade into obscurity: While theoretical it is not loosely speculative; while dealing with uncommon and sometimes enigmatic observations it avoids exoticisms; and it provides a substantial basis for the practical application of its wealth of information about the class of neuropathological conditions that can best be understood as subcortical dementias. Thus Dr. Derix's well-researched dissertation could evolve into a book that combines a theoretically sound discussion of the concept and kinds of subcortical dementia with a practical clinical approach which is readily accessible to practitioners and students of clinical neurosciences.

I believe that this book is the first to review and synthesize the neuropathology, cognitive theory, and practical neuropsychological assessment aspects of all the major subcortical dementias – and to make the necessary discriminations between these various similar but quite different progressively deteriorating processes – from a neuropsychological viewpoint. Neuropsychological practitioners performing evaluations will find the clinically well-grounded information in this book useful. Equally useful is this same information for neurologists, speech pathologists, psychiatrists, and other clinical specialists who need to work knowledgeably with neuropsychologists to provide the best care for these patients. Dr. Derix has done neuropsychology a valuable service and, in so doing, has served these patients and their clinicians well.

Muriel D. Lezak, Ph.D.
Professor, Neurology, Psychiatry & Neurosurgery
Oregon Health Sciences University

Contents

Acknowledgement

It is a pleasure to be able to express my thanks to those who have given their help and support during the preparation of my thesis and finally this book.

Professor dr. H. van Crevel created the environment in which neuropsychology became an integral part of patient care and clinical research in the department of neurology in the Academic Medical Centre in Amsterdam. Dear Hans, you stimulated research into the field of dementia. Thank you for guidance and critical appraisal of my work during my stay at the Academic Medical Centre, which ultimately led to the completion of this book.

Special thanks go to dr. Albert Hijdra, dr. Jaap Lindeboom, drs. Erny Groet, dr. Peter Portegies, and dr. Jan de Gans.

Dear Albert, your interest in clinical neuropsychology and differentiation between dementia syndromes made it a pleasure to discuss and write about this subject with you. Thank you for your support, critical questions and remarks.

Dear Jaap, you made me feel at home within the field of neuropsychology and I thank you for your practical advice, and your critical comments on my clinical and scientific endeavour.

Dear Erny, during your stay at the Academic Medical Centre the idea of neuropsychological differentiation between dementia syndromes gained a firm footing and resulted in a first publication on this subject.

Dear Peter and Jan, thank you for your discussions about the mental changes in patients with AIDS, and pleasant cooperation in writing about the different findings.

Very special thanks go to my colleague drs. Saskia Teunisse. Dear Saskia, our mutual interest in dementia, our discussions about various neuropsychological subjects and the assessment of the severity of dementia, have made the completion of this book a worthwhile goal.

I thank my parents for their support, and for offering a quiet and healthy surrounding when energy levels became low.

I wish to thank professor Muriel Lezak for her suggestion to try to get a shortened version of my thesis published, and dr. Paul Dijkstra from Swets & Zeitlinger BV for making this possible.

Last, but certainly not in the least, I express my thanks to my husband. Dear Guus, your unfailing support, encouragement, humour, computerized and mobile assistance made the completion of this book possible.

CHAPTER 1

Introduction

In recent years the number of publications related to the study of patients with (possible) dementia has grown substantially. This can be partly attributed to the growing number of elderly people in the population and the related increase of the number of patients with mental deterioration, but also to the improvement of diagnostic possibilities.

About 5% to 10% of people above 65 years will show mental impairment, with an increase to 20% to 25% above 85 years. But global mental deterioration can also occur in younger age groups. It is estimated that 50% of patients with dementia suffers from Alzheimer's disease. In 10% to 20% dementia is caused by cerebrovascular disorders, with various manifestations [17]. In another 10% to 20% a combination of Alzheimer's disease and cerebrovascular disorders is thought to be the cause. But there are many other diseases which can lead to or are accompanied by dementia [1,23-25,28,29]. In the differentiation between the various causes the distinction between cortical and subcortical dementia can be useful [8].

The majority of patients with dementia suffers from an irreversible condition, but there are also patients with a stationary or a reversible condition. Careful, thorough diagnostic examination is pertinent for detection or exclusion of those disorders [5-7]. The same requirement has to be fulfilled to outline the various behavioural alterations and to predict the course of the disease, which is necessary for giving information, advice and guidance to family members, physicians responsible for the management of these patients, and institutional staff. Not only the patient's life but also that of his family is affected by the behavioural alterations due to brain damage [21].

In the diagnostic examination of patients with (possible) dementia attention must be paid to both physical and mental functions. With regard to the second aspect, neuropsychological evaluation of the cognitive functions of a patient can be a valuable contribution, as the assessment of mental state is probably the single most important phase in the diagnosis of dementia.

Neuropsychology is a relatively new scientific discipline, directed at the study of brain-behaviour relationships [20]. It evolved mainly out of psychology, although it can also be seen as a compound discipline, because it represents the confluence of several other fields of study, e.g. neurology, neuroanatomy, neurophysiology, neurochemistry and neuropharmacology [4,20]. The results of neuropsychological studies have led to a better understanding of different aspects of normal and impaired behaviour. Behaviour denotes a general concept and includes functions such as language, writing, reading, arithmetic, visuospatial skill, attention, memory, reasoning and problem solving. Clinical neuropsychology is directed at the examination of patients with cognitive impairments, caused by disorders of the brain. Various tests, rating scales and questionnaires are used. The goal of neuropsychological studies involves the qualitative and quantitative analysis of the behavioural sequela of brain damage. The strength of a neuropsychological examination is that it can provide a profile of the impaired and also of the intact cognitive functions. These profiles can be used in diagnosis, patient care, and research [12].

In evaluating the older person, neuropsychology is challenged more than by any other age group. Because of methodological shortcomings (e.g. a lack of normative test data), and non-methodological shortcomings (e.g. neglect of motivational factors, of the effects of fatigue and the unusualness of test situations) one has to be extra careful in interpreting the results of neuropsychological studies performed in this age group [3,12,16,19].

A promising development has been the introduction of the concepts of cognitive psychology in clinical psychology and in neuropsychology in particular [18,26]. Cognitive neuropsychology seeks to explain the patterns of impaired and intact cognitive performance in terms of damage to one or more of the components of a model of normal cognitive functioning [15]. The question for modern clinical neuropsychology is not how to localize cerebral lesions, but rather how to consistently relate neuropsychological dysfunctions and functions with findings from other investigations, for instance neuroimaging results. The next step is to explain the behaviour of the patient in the light of these findings and in terms of a cognitive model of functioning.

From the end of the 19th century, cognitive impairment has been a recognized part not only of cortical but also of subcortical disorders. In 1932 the term subcortical dementia was used to describe the mental deterioration occurring in postencephalitic Parkinson's disease [22]. Since the mid-1970's there has been an increasing awareness that diverse degenerative brain disorders may lead to mental deterioration. It became clear that there were clusters of behavioural and cognitive changes which differed from that of Alzheimer's disease [22]. In 1974 Albert et al [2] reintroduced the concept of subcortical dementia in a study of progressive supranuclear palsy. They contrasted it with cortical dementia of which Alzheimer's disease is the prototypical example. From then on, the number of neurological diseases with the clinical characteristics of a subcortical type of dementia has expanded considerably [8,27]. The distinction between subgroups of dementia syndromes is especially made on the basis of behavioural-neurological and neuropsychological findings.

This book is an adaptation of a doctoral thesis [9]. The results of neuropsychological studies of patient groups with various dementia syndromes, published elsewhere [10,11,13,14], are incorporated in a chapter on the possibility and usefulness of the distinction of various dementia syndromes. The main focus is on the manifestations of various subcortical dementia syndromes contrasted with cortical dementia, with Alzheimer's disease as the main example.

The results of an extensive literature review with regard to dementia in diseases predominantly affecting subcortical structures will be reported. These include progressive supranuclear palsy, Parkinson's disease, Huntington's disease, normal pressure hydrocephalus, subcortical arteriosclerotic encephalopathy, Aids Dementia Complex, and multiple sclerosis. First, the clinical manifestations, pathology and pathophysiology will be described, followed by a literature review of results of studies concerning performance on neuropsychological tasks. An attempt will be made to explain the reported cognitive deterioration in 'subcortical' and also in 'cortical' dementia in terms of a cognitive, neuropsychological model of information processing. In this model the function of the frontal lobes plays a prominent part.

To illustrate the usefulness of clinical neuropsychological examination of patients with signs and symptoms of various types of dementia, three case histories will be presented. Guidelines will be offered for neuropsychological assessment in dementia. Information will be given with regard to the principal characteristics of various dementia syndromes and the problems encountered in neuropsychological evaluation of dementia.

References

1. Advies Gezondheidsraad nr 1988/07. (Recommendation of the Council of Health, nr. 1988/07)
   Psychogeriatrische ziektebeelden. (Psychogeriatric syndromes).
   's-Gravenhage, 1988 (The Hague, 1988)

2. Albert ML, Feldman RG, Willis AL.
   The 'subcortical dementia' of progressive supranuclear palsy.
   J Neurol Neurosurg Psychiat 1974; 37: 121-30

3. Albert MS.
   Geriatric Neuropsychology.
   J Consult Clin Psychol 1981; 49: 835-50

4. Benton A.
   Neuropsychology: past, present and future.
   In: Boller F, Grafman J (Eds). Handbook of Neuropsychology, vol. 1.
   Amsterdam, Elsevier Science Publ BV, 1988: 1-27

5. Clarfield A.
   The reversible dementias: do they reverse?
   Ann Int Med 1988; 109: 476-86

6. Crevel H van, Teunisse S, Otten JMMB.
    Reversibele dementieën: meten van behandelingseffecten. (Reversible
    dementias: measuring the effect of treatment.)
    Tijdschr Gerontol Geriatr 1989; 20: 249-50

7. Crevel H van.
    Clinical approach to dementia.
    In: Swaab DF, Fliers E, Mirmiran M, et al (Eds). Aging of the brain and
    senile dementia. Progress in Brain Research 70.
    Amsterdam, Elsevier Science Publ BV, 1986; 70: 3-13

8. Cummings JL, Benson DF.
    Dementia: a clinical approach. sec ed.
    Boston, Butterworth, 1991

9. Derix MMA.
    Neuropsychological differentiation of dementia syndromes. Thesis.
    Amsterdam, University of Amsterdam, 1991

10. Derix MMA, de Gans J, Stam J, Portegies P.
    Mental changes in patients with AIDS.
    Clin Neurol Neurosurg 1990; 92: 215-22

11. Derix MMA, Groet E.
    Subcorticale en corticale dementie: een neuropsychologisch onder-
    scheid. (Cortical and subcortical dementia: a neuropsychological
    distinction.)
    In: Schroots JJF, Bouma A, Braam GPA et al (Eds). Gezond zijn is
    ouder worden.
    Assen, Van Gorcum, 1989: 179-88

12. Derix MMA, Teunisse S.
    Neuropsychologisch onderzoek bij ouderen. (Neuropsychological
    assessment of the elderly.)
    Bijblijven 1989; 5: 16-23

13. Derix MMA.
    Corticale en subcorticale dementie: een zinvol onderscheid. (Cortical
    and subcortical dementia: a useful distinction?)
    Ned Tijdschr Geneesk 1987: 131: 1070-4

14. Derix MMA, Hijdra A, Verbeeten BWJ jr.
    Mental changes in subcortical arteriosclerotic encephalopathy.
    Clin Neurol Neurosurg 1987; 89: 71-8

15. Ellis AW, Young AW.
    Human cognitive neuropsychology.
    London, Lawrence Erlbaum Associates, Publishers, 1988

16. Gallagher D, Thompson LW, Levy SM.
    Clinical assessment of older adults.
    In: Poon LW (Ed). Ageing in the 1980s, psychological issues.
    Washington, American Psychological Association Inc, 1980: 19-40

17. Hijdra A.
    Vascular dementia.
    In: Bradley WG, Daroff RB, Fenichel GM, Marsden CD (Eds).
    Neurology in clinical practice.
    Boston, Butterworth, 1990: 1425-35

18. Ingram RE, Kendall PC.
    Cognitive clinical psychology: implications of an information processing
    perspective.
    In: Ingram RE (Ed). Information processing approaches to clinical
    psychology.
    New York, Academic Press Inc., 1986: 3-21

19. Kaszniak AW.
    Neuropsychological consultation to geriatricians: issues in the
    assessment of memory complaints.
    The Clin Neuropsychologist 1987; 1: 35-46

20. Lezak MD.Neuropsychological assessment, 2nd ed. (3rd ed. in press,
    1994)
    New York, Oxford University Press, 1983

21. Lezak MD.
    Brain damage is a family affair.
    J Clin Exp Neuropsychol 1988; 10: 111-23

22. Mandell AM, Albert ML.
    History of subcortical dementia.
    In: Cummings JL (Ed). Subcortical dementia.
    Oxford, Oxford University Press, 1990: 17-30

23. Rossor M.
Dementia as part of other degenerative diseases.
In: Bradley WG, Daroff RB, Fenichel GM, Marsden CD (Eds).
Neurology in clinical practice.
Boston, Butterworth, 1990: 1423-24

24. Rossor M.
Miscellaneous causes of dementia
In: Bradley WG, Daroff RB, Fenichel GM, Marsden CD (Eds).
Neurology in clinical practice.
Boston, Butterworth, 1990: 1436-41

25. Schoenberg BS.
Epidemiology of dementia.
In: Hutton JT (Ed). Neurologic Clinics. Vol 4 (2): Dementia.
Philadelphia, WB Saunders Company, 1986: 447-57

26. Shallice T.
From neuropsychology to mental structure.
Cambridge, Cambridge University Press, 1988

27. Subcortical dementia.
Cummings JL (Ed).
Oxford, Oxford University Press, 1990

28. Tomlinson BE, Blessed G, Roth M.
Observations on the brains of demented old people.
J Neurol Sci 1970; 11: 205-42

29. Wells CE.
Diagnostic evaluation and treatment in dementia.
In: Wells CE (Ed). Dementia.
Philadelphia, FA Davis, 1977: 247-76.

CHAPTER 2

Cortical and subcortical dementia
A possible and useful distinction?

Introduction

In the neurological and neuropsychological literature from the past ten years much attention has been paid to the various cognitive disorders caused by subcortical pathology. The discussion often centres on possible cognitive differences in comparable cortical and subcortical pathology. Neuropsychological examination can reveal differences between localized cortical and subcortical lesions [27]. However, many patients referred for neuropsychological examination often present with signs and symptoms of an extensive diffuse cerebral disease. This may be a cortical or subcortical progressive degenerative process, or a combination of the two.

Dementia is often associated with degenerative disorders of the cerebral cortex, of which Alzheimer's disease is the best known example. Dementia may, however, also be the result of disorders which mainly affect subcortical structures or conditions with both cortical and subcortical involvement. Clinical differentiation of the dementia syndrome in cortical, subcortical and mixed types has been emphasized [1,15,16,61]. The nature of the differences between these dementia syndromes is still not resolved [45,66]. The distinction however has stimulated investigation into the field of dementia. It also may have clinical significance because important behavioural and neuropsychological differences are stressed, which could be helpful in diagnosis and possible also in advising the carers of dementia patients[15].

Definition and classification of dementia

The definition of dementia is still controversial. Among other things this is due to different disciplines, each emphasizing different aspects of the concept [53]. Dementia is most commonly defined as a clinical syndrome of deterioration of mental functioning, in the absence of disturbed consciousness, with impairment in at least three or more of the following domains: memory, visuospatial skills, personality, language and other acquired abilities such as arithmetic, reading, writing, judgement and abstract reasoning [15,38]. Personality changes can be present and often it is added that this deterioration should interfere with social and/or professional functioning [23]. In the scope of this definition dementia may be static (e.g. posttraumatic and postencephalitic encephalopathy), progressive (e.g. Alzheimer's disease), but also reversible (e.g. normal pressure hydrocephalus) [42]. As a rule, dementia can be considered a complex of signs and symptoms which may occur in many diseases [32].

The general description of dementia seems to allows for the distinction of two principal subgroups on the basis of behavioural-neurological and neuropsychological findings: cortical and subcortical dementia. Dementia with characteristics of both types also occurs. The classic example of cortical dementia is that in Alzheimer patients. The best known examples of subcortical dementia may be found in Parkinson's disease, Huntington's disease and progressive supranuclear palsy (also called the Steele-Richardson-Olzewski syndrome). It has been shown that there are clear differences with regard to results on neuropsychological evaluation between patients with Alzheimer's disease compared with mentally deteriorated patients with Parkinson's disease, Huntington's disease and with progressive supranuclear palsy [12,20,49,54]. The distinction is mainly possible in the early stages of these diseases and is often not solely based on psychometric quantitative aspects of the test data: evaluation of quantitative aspects must also be taken into account [20,36]. The later stages of the underlying disorders may show strong neuropathological similarity with respect to localization [6,66], and a clinical distinction will eventually become extremely difficult if not impossible. The distinction may be of great clinical importance, because most treatable disorders are accompanied by dementia of a subcortical type [15]. A detailed neuropsychological examination can be a valuable ancillary investigation in the differentiation between dementia syndromes [21].

Cortical dementia

The dementia in Alzheimer's disease is the best known example of cortical dementia. Alzheimer's disease is a cerebral degenerative process, accompanied by progressive mental deterioration. The pathological changes mainly affect the association areas of the parietal, temporal and frontal lobes and the hippocampus [4]. In the early stages mainly cortical dysfunctions attract attention, particularly of the parietal and posterior temporal regions, as can be shown by studies of cerebral blood flow and metabolism [8,13,17,35]. In these stages the cognitive decline is characterized by memory deficits, impaired judgement and abstraction ability, aphasia, acalculia, visuospatial defects and personality changes, such as behavioural disinhibition and diminishing interest. Speech and motor function remain unimpaired. In the later stages specific cognitive defects, such as aphasia, apraxia and agnosia, become progressively more severe [3,9,15,37,40,47,60]. Distinct subgroups of patients with Alzheimer's disease have been reported, characterized qualitatively by different profiles of cognitive impairment and corresponding patterns of cerebral hypometabolism [43,52]. Pick's disease also leads to cortical dementia. Although the clinical picture and course of the dementia in this disease can be very similar to those in Alzheimer's disease, accurate observation and examination does often show differences in the clinical picture, in accordance with neuropathological differences [15,53,57,62,63].

In Alzheimer's disease the pathological changes are mainly concentrated in the association areas of the parietal, temporal and frontal lobes and in the hippocampus. In Pick's disease there are mainly changes in the frontotemporal and frontal areas. It seems plausible to assume that this difference would make the process of becoming demented different. There is a survey of the behavioural-neurological characteristics which might make it possible to differentiate between these diseases [15]. The value of these character-istics in the differential diagnosis is dependent on the time of occurrence in the course of the illness. For instance, amnesia and spatial disorientation occur at an early stage in Alzheimer's disease; in Pick's disease personality and behavioural changes are predomi-nant in the initial stages. Even in patients with the same disease, however, there may be individual differences of symptoms [25,44].

Subcortical dementia

This concept was introduced in 1974 as a result of an investigation of the mental functioning of patients with progressive supranuclear palsy [1]. One year later a corresponding dementia syndrome was described in patients with Huntington's disease [46].

The clinical syndrome is characterized by slowing of cogni-tion, mnestic dysfunction and difficulty with complex intellectual tasks requiring strategy generation and problem solving. The memory impairment, defined as 'forgetfulness', consists of a de-fect in the recall of already acquired information. Changes of affect are often present. Aphasia, apraxia and agnosia are absent in subcortical dementia syndromes [15,20]. Motor impairment is a frequent feature (for instance rigidity or a pseudobulbar syn-drome), with involvement of speech (hypophonia and (or) dys-arthria) [15,16].

Subcortical dementia may occur in a large number of dis-eases such as Parkinson's disease, hydrocephalus, multiple scle-rosis, toxic and metabolic encephalopathies, Huntington's disease, progressive supranuclear palsy, spinocerebellar degener-ations and Wilson's disease [15,16,40,42]. Even in vascular dementia such as in a lacunar state and in Binswanger's encephalopathy, a subcortical form of dementia has been described [16,22,41]. The AIDS Dementia Complex (ADC), a dementia syndrome that can occur in the course of the Acquired Immunodeficiency Syndrome (AIDS), also has the characteristic features of a subcortical type of dementia [19,51]. Finally, it has been put forward that the changes in mental functioning which are often present in severely depressive patients, are similar to those of a subcortical dementia syndrome [11].

It is plausible to assume that within the general description of the clinical syndrome of subcortical dementia, subgroups can be distinguished on the basis of the different localizations of the subcortical lesions. In a study, patients with Huntington's disease and patients with multiple sclerosis, all having cognitive impairment, were compared as to memory, language, visuospatial and arithmetical skills. The performances in the two groups had a global similarity and were different from those in a control group. The groups however could be distinguished on account of memory and arithmetical impairment, which were more severe in patients with Huntington's disease than in patients with multiple sclerosis [10]. Subtle differences have also been found between patients with Parkinson's disease and patients with progressive supranuclear palsy. The latter group exhibits more frontal-type behaviour, performs slightly worse on so-called frontal lobe tests, and is more impaired on tasks measuring cognitive slowing [24,55].

Mixed dementia

Of course, there are also dementia syndromes with both cortical and subcortical features. Examples are multi-infarct dementia, dementia in neurosyphilis and the dementia in postanoxic and posttraumatic encephalopathies [16].

There is also a fourth form of dementia, namely axial dementia [53]. Pathologically this concerns disorders of axial structures such as hippocampus, fornix, mammillary bodies, hypothalamus and the medial part of the temporal lobe. The classic example is Korsakoff's syndrome. This syndrome, however, does not meet the criteria of the diagnosis of dementia (see above), because the neuropsychological impairment mainly concerns the patient's memory. The past six years however many studies with regard to neurobehavioural and neuropsychological examination of chronic alcoholics have mentioned a subgroup which appears to have a (Wernicke-)Korsakoff syndrome but shows more cognitive dysfunction than amnesia alone. Often the presence of two or more of the following cognitive deficits are mentioned: poor abstracting abilities, impairment of verbal fluency, disorientation, poor attention, the occurrence of perseverations, psychomotor retardation, and mood changes. These patients may therefore be diagnosed as suffering from 'dementia associated with alcoholism'[64].

Pathophysiological background

Although the so-called higher cerebral functions are traditionally associated with the cerebral cortex, a great number of subcortical structures are strongly involved in these functions. Cortico-cortical connections are of great importance for the integration of cortical processes, and these connections are localized for a great part in the subcortical white matter. Focal lesions in the white matter may result in so-called disconnection syndromes [28]. An example is conduction aphasia with a lesion of the arcuate fasciculus, which connects the cortical areas of Wernicke and Broca. Hippocampus and adjacent cortex, fornix, mammillary bodies, amygdala, and the dorsomedial nucleus of the thalamus play an important part in memory and related functions [18,58,59]. Bilateral lesions in these structures, especially in the so-called medial temporal lobe system, result in memory disorders (for instance Korsakoff's syndrome). The reticular formation, the raphe nuclei and the locus coeruleus are essential for the activation of the cortex, directly or through specific and aspecific nuclei of the thalamus, and in directing and keeping attention onto outside stimuli.

Thus, some subcortical lesions may cause impairments, which may superficially resemble those of cortical origin, for instance speech and language disturbances with lesions of the white matter and of the thalamus [30,34]. Other subcortical lesions may influence cortical functions in a more general way by impairment of activation, attention and temporal integration with reference to all cortical functions [48].

This connection between cortex and subcortical structures, has been known for a long time, both neuroanatomically and neurophysiologically. It has been elegantly demonstrated with positron emission tomography, which can visualize hypometabolic cortical areas during life in patients with local lesions of the thalamus [5]. In patients with progressive supranuclear palsy this method could show decreased metabolism in both frontal lobes, which are not affected anatomically in this disorder [2]. This hypometabolism must be explained by a de-afferentiation of the cortical areas involved by lesions in subcortical areas.

Arguments pro and contra the distinction

Purely cortical and subcortical degenerative diseases do not really exist. In Parkinson's disease there are also some cortical and in Alzheimer's disease some subcortical pathological changes (for instance in the basal nucleus of Meynert) [33,65]. In the later stages of degenerative diseases the differentiation on the basis of anatomical localizations becomes difficult. Clinically these disorders become more and more similar (in Alzheimer patients for instance hypokinetic rigidity will also develop). Perhaps it was an unfortunate decision to make a clinical and neuropsychological distinction between dementia syndromes on the basis of anatomical structures, but this semantic problem does not have to make the distinction less valuable.

One of the most important features of the difference between cortical and subcortical dementia is the presence or absence of aphasia. In focal subcortical lesions, language disturbances may occur, but they can be distinguished from the classic cortical aphasias. A remarkable fact is that subcortical aphasia often recovers well in spite of a permanent subcortical structural lesion [7]. Even in patients with Parkinson's disease who had undergone a thalamotomy in the dominant hemisphere for the treatment of tremor, language disturbances were found after the operation, which, however, could not be ranged under the known aphasia syndromes [31]. Comprehension was intact, there were no disturbances in naming and no grammatical errors, and repetition was also normal. The language disturbances were mainly characterized by word-finding problems and a reduced word fluency. A few months after the operation the patients only showed a reduced word fluency and word-finding problems in spontaneous speech.

Much investigation into the degree of mental deterioration has been done with short screening tasks such as the Mini Mental

State Examination (MMSE). In a study minimal and mild dementia could be detected in patients with probable Alzheimer's disease, but demented patients with Parkinson's disease and with progressive supranuclear palsy reached scores above the criteria, in spite of their showing clear signs of mental deterioration on extensive assessment [20]. An other choice has often been the use of extensive tests such as the Wechsler Adult Intelligence Scale, the Halstead-Reitan and the Wechsler Memory Scale. These are an intelligence test, a test to investigate the presence of brain lesions and a memory test, respectively. These tests consist of a great number of subtests, but the final result is a total score, in which much valuable information is lost. In such an examination it often appears that on account of these quantitative test results (total scores) it is not possible to find differences between groups of demented patients [45]. The problem with these tests is that both patients with a cortical and those with a subcortical dementia will fail the tests. In this way, however, the different backgrounds of this failure do not come to light [16].

In neuropsychological examination directed at the qualitative differences between dementia syndromes, these reasons do emerge. With the help of seven tasks a neuropsychological difference could be shown between patients with Alzheimer's disease and with Parkinson's disease [36]. The latter group was characterized by good orientation for time, place and person, and by absence of aphasia and apraxia. In both groups memory and visuospatial disturbances were present. Comparable results were found by Derix and Groet (1989) [20]. Other authors examined the memory of, among others, patients with Alzheimer's or Huntington's disease, and found a marked difference [50]. Alzheimer patients showed a marked learning impairment, whereas Huntington patients, in verbal memory tests, showed disturbances in retrieving information from memory.

It has not been sufficiently investigated whether the neuropsychological difference on account of which the two dementia syndromes have been defined, is supported by differences on other, for instance neurophysiological grounds. From some studies it appears that the latency times of certain components of auditory evoked cortical responses differ significantly between Alzheimer patients on the one hand and dementing patients with Parkinson's or Huntington's disease on the other [29]. Unfortunately, neuropsychological differences between the two groups of patients were not studied. Further investigation in this direction is necessary, and is interesting from a clinical as well as a pathophysiological point of view.

Clinical significance of the distinction

The distinction between cortical and subcortical dementia has had a stimulating effect on the thinking about dementia and on the investigation of mechanisms and causes of dementia syndromes. Thus, on the basis of the described clinical differences and the increasing neuroanatomical and neurophysiological knowledge, a classification of dementias based on neurotransmitter deficits may become possible, from which pharmacological treatment could be developed.

It is of practical significance that most treatable disorders causing dementia (for instance hydrocephalus and toxic and metabolic encephalopathies) present with a subcortical type of dementia. An exception can be found in patients with AIDS. A cortical type of dementia in patients with AIDS can be explained by the presence of occasionally treatable opportunistic infections [19]. Also the incidence of the Aids Dementia Complex (ADC), which is of a subcortical type, has been shown to decline strikingly after the introduction of antiviral treatment (AZT - Zidovudine) [56]. Thus, on clinical and neuropsychological grounds it could be decided what ancillary investigations are indicated in individual patients [14].

When advising members of a patient's family, the distinction may also be important. What goes wrong in thinking and in the behaviour in a patient with a subcortical dementia syndrome is very much determined by the period of time given to him and the degree of activation. Activation and patience will not much improve the behavioural disturbances caused by amnesia, spatial disorientation, aphasia and the like in a patient with cortical dementia.

Conclusion

The distinction between cortical and subcortical dementia seems, in spite of the drawbacks attached to it, a useful asset in thinking about dementia and for the clinical practice with regard to diagnosis, treatment and attendance to demented patients. Especially, in decisions about ancillary investigations that are necessary for patients, the distinction may be of practical use. It is important to know that the difference will show in neuropsychological examination, particularly with qualitative analysis of the results. With a psychometric approach only, valuable information may be lost. The strict handling of time limits is to the disadvantage of patients with a subcortical dementia. It is not only the question of dementia being present or not that has to be answered, but also which are its main characteristics.

References

1. Albert ML, Feldman RG, Willis AL.
    The 'subcortical dementia' of progressive supranuclear palsy.
    J Neurol Neurosurg Psychiat 1974; 37: 121-30

2. Antona R d', Baron JC, Samson Y, et al.
    Subcortical dementia. Frontal cortex hypometabolism detected by
    positron tomography in patients with supranuclear palsy.
    Brain 1985; 108: 785-99

3. Appell J, Kertesz A, Fisman M.
    A study of language functioning in Alzheimer patients.
    Brain and Language 1982; 17: 73-91

4. Arnold SE, Hyman BT, Flory J, Damasio AR, Van Hoesen GW.
    The topographical and neuroanatomical distribution of neurofibrillary
    tangles and neuritic plaques in the cerebral cortex of patients with
    Alzheimer's disease.
    Cerebral Cortex 1991; 1: 103-16

5. Baron JC, Antona R d', Serdaru M, et al.
   Hypometabolisme cortical après lesion thalamique chez homme: étude
   par la tomographie á positrons.
   Rev Neurol 1986; 142: 465-74

6. Benson DF.
   Parkinsonian dementia: cortical or subcortical?
   In: Hassler RG, Christ JF (Eds). Parkinson-specific motor
   and mental disorders. Adv Neurol 40.
   New York, Raven Press, 1984: 235-40

7. Benson DF, Geschwind N.
   Aphasia and related disorders: a clinical approach.
   In: Mesulam M-M (Ed). Principles of behavioral neurology.
   Philadelphia, FA Davis Company, 1985: 193-238

8. Berg G, Grady Cl, Sundaram M, et al.
   Positron emission tomography in dementia of the Alzheimer type.
   Arch Intern Med 1986; 146: 2045-49

9. Botwinick J, Storandt M, Berg LA.
   A longitudinal, behavioral study of senile dementia of the Alzheimer
   type.
   Arch Neurol 1986; 43: 1124-7

10. Caine ED, Bamford KA, Schiffer RB, et al.
    A controlled neuropsychological comparison of Huntington's disease
    and multiple sclerosis.
    Arch Neurol 1986; 43: 249-54

11. Caine ED.
    Pseudodementia.
    Arch Gen Psychiatry 1981; 38: 1359-64

12. Caltagirone C, Carlesimo A, Nocentini U, Vicari S
    Differential aspects of cognitive impairment in patients suffering from
    Parkinson's and Alzheimer's disease: a neuropsychological evaluation.
    Int J Neurosci 1989; 44: 1-7

13. Chase TN.
Focal cortical abnormalities in Alzheimer's disease as determined by positron emission tomography.
In: Wurtman RJ, Corkin SH, Crowdon JH (Eds). Alzheimer's disease: advances in basic research and therapies.
Proceedings of the third meeting of the international study group on the treatment of memory disorders associated with aging.
Zürich, Switzerland, january 13-15, 1984: 95-110

14. Crevel H van.
Clinical approach to dementia.
In: Swaab DF, Fliers E, et al. (Eds). Aging of the brain and senile dementia. Progress in brain research 70.
Amsterdam, Elsevier Science Publ BV, 1986: 3-13

15. Cummings JL, Benson DF.
Dementia: a clinical approach. sec ed.
Boston, Butterworth, 1992

16. Cummings JL, Benson DF.
Subcortical dementia: review of an emerging concept.
Arch Neurol 1984; 41: 874-97

17. Cutler NR, Haxby JV, Duara R, et al.
Clinical history, brain metabolism, and neuropsychological function in Alzheimer's disease.
Ann Neurol 1985; 18: 289-309

18. Derix MMA.
Geheugen, dementie en de hersenen. (Memory, dementia and the brain.)
In: van Crevel H. et al: Perspectief in de neurologie.
Houten, Bohn Stafleu Van Loghum, 1992: 25-36

19. Derix MMA, de Gans J, Stam J, Portegies P.
Mental changes in patients with AIDS.
Clin Neurol Neurosurg 1990; 92: 215-22

20. Derix MMA, Groet E.
Subcorticale en corticale dementie: een neuropsychologisch onderscheid. (Subcortical dementia: a neuropsychological distinction.)In: Schroots JJF, Bouma A, Braam GPA et al (Eds).
Gezond zijn is ouder worden.
Assen, Van Gorcum, 1989: 179-88

21. Derix MMA, Hijdra A.
    Corticale en subcorticale dementie: een zinvol onderscheid? (Cortical and subcortical dementia: a useful distinction?)
    Ned Tijdschr Geneesk 1987; 131: 1070-4

22. Derix MMA, Hijdra A, Verbeeten B.
    Mental changes in subcortical arteriosclerotic encephalopathy.
    Clin Neurol Neurosurg 1987; 89: 71-8

23. DSM-III-R. Diagnostic and statistic manual of mental disorders, 3rd ed. revised.
    Washington DC, American Psychiatric Association, 1987

24. Dubois B, Pillon B, Legault F, et al.
    Slowing of cognitive processing in progressive supranuclear palsy.
    Arch Neurol 1988; 45: 1194-9

25. Filley CM, Kelly H, Heaton RK.
    Neuropsychological features of early- and late-onset Alzheimer's disease.
    Arch Neurol 1986; 43: 574-6.

26. Friedland RP, Jagust WJ, Huesman RH et al.
    Regional cerebral glucose transport and utilization in Alzheimer's disease.
    Neurol 1989; 39: 1427-34

27. Fromm D, Holland AL, Swindell CS, Reinmuth OM.
    Various consequences of subcortical stroke.
    Arch Neurol 1985; 42: 943-50

28. Geschwind N.
    Disconnection syndromes in animals and man.
    Brain 1965; 88: 237-94

29. Goodin DS, Aminoff MJ.
    Electrophysiological differences between subgroups of dementia.
    Brain 1986; 109: 1103-13

30. Graff-Radford NR, Damasio H, Yamada T, Eslinger PJ, Damasio AR.
    Nonhaemorrhagic thalamic infarction: clinical, neuropsychological and electrophysiological findings in four anatomical groups defined by computerized tomography.
    Brain 1985; 108: 485-516

31. Groet E.
Strubbelingen tussen mijn hand en mijn verstand. Cognitieve defecten
na een thalamotomie. (Troubles between my hand and mind. Cognitive
deficits after thalamotomy.)
Amsterdam, Universiteit van Amsterdam, Vakgroep Psychologie, 1985.
(Amsterdam, University of Amsterdam, Dept. Psychology, 1985)

32. Haase GR.
Diseases presenting as dementia.
In: Wells CE (Ed). Dementia, 2nd ed.
Philadelphia, FA Davis Company, 1977: 27-68

33. Hakim AM, Mathieson G.
Dementia in Parkinson's disease: a neuropathologic study.
Neurology 1979; 29: 1209-14

34. Herderschee D, Stam J, Derix MMA.
Aphemia as a first symptom of multiple sclerosis.
J Neurol Neurosurg Psychiatr 1987; 50: 499-50

35. Hoffman JM, Guze BH, Baxter LR, et al.
(^{18}F)-Fluorodeoxyglucose (FDG) and positron emission tomography
(PET) in aging and dementia.
Eur Neurol 1989; 29 (suppl 3): 16-24

36. Huber SJ, Shuttleworth EC, Paulson GW et al.
Cortical versus subcortical dementia: neuropsychological differences.
Arch Neurol 1986; 43: 392-4

37 Huff FJ, Corkin S, Growdon JH.
Semantic impairment and anomia in Alzheimer's disease.
Brain and Language 1986;28:235-49

38. Joynt RJ, Shoulson I.
Dementia.
In: Heilman KM, Valenstein E (Eds). Clinical neuropsychology.
New York, Oxford University Press, 1979; 475-502

39. Katzman R.
Alzheimer's disease.
N Engl J Med 1986; 314: 964-72

40. Katzman R.
Differential diagnosis of dementing illnesses.
Neurol Clinics 1986; 4: 329-40

41. Kinkel WR, Jacobs L, Polachini I, et al.
Subcortical arteriosclerotic encephalopathy (Binswanger's disease):
computed tomographic, nuclear magnetic resonance, and clinical
correlations.
Arch Neurol 1985; 42: 951-9

42. Kirshner HS.
Behavioral neurology: a practical approach.
New York, Churchill Livingstone, 1986: 143-79

43. Martin A, Brouwers P, Lalonde F et al.
Towards a behavioral typology of Alzheimer's patients.
J Clin Experim Neuropsychol 1986; 8: 594-610

44. Mayeux R, Stern Y, Spanton S.
Heterogeneity in dementia of the Alzheimer type: evidence of sub-
groups.
Neurol 1985; 35: 453-61

45. Mayeux R, Stern Y, Benson DF.
Is 'subcortical dementia' a recognizable entity?
Ann Neurol 1983; 14: 278-83

46. McHugh PR, Folstein MF.
Psychiatric syndromes of Huntington's chorea: a clinical and phenome-
nologic study.
In: Benson DF, Blumer D (Eds). Psychiatric aspects of neurologic
disease.
New York, Grune and Stratton, 1975: 267-85

47. McKhann G, Drachman D, Folstein M et al.
Clinical diagnosis of Alzheimer's disease: Report of the NINCDS-
ADRDA Work Group under the auspices of Department of Health and
Human Services Task Force on Alzheimer's disease.
Neurol 1984; 34: 939-44

48. Mesulam M-M.
    Patterns of behavioral neuroanatomy: association areas, the limbic
    system, and hemispheric specialization.
    In: Mesulam M-M (Ed). Principles of behavioral neurology.
    Philadelphia, FA Davis Company, 1985: 1-70

49. Milberg W, Albert W.
    Cognitive differences between patients with progressive supranuclear
    palsy and Alzheimer's disease.
    J Clin Exp Neuropsychol 1989; 11: 605-14

50. Moss MB, Albert MS, Butters M, Payne M.
    Differential patterns of memory loss among patients with Alzheimer's
    disease, Huntington's disease, and alcoholic Korsakow syndrome.
    Arch Neurol 1986; 43: 239-46

51. Navia BA, Jordan BD, Price RW.
    The AIDS Dementia Complex: I. Clinical features.
    Ann Neurol 1986; 19: 517-24

52. Neary D, Snowden JS, Bowen DM, Sims NR et al.
    Neuropsychological syndromes in presenile dementia due to cerebral
    atrophy.
    J Neurol Neurosurg Psychiatr 1986; 49: 163-74

53. Pearce JMS.
    Dementia: a clinical approach.
    Oxford, Blackwell Scientific Publications, 1984

54. Pillon B, Dubois B, Ploska A, Agid Y.
    Severity and specificity of cognitive impairment in Alzheimer's,
    Huntington's, and Parkinson's diseases and progressive supranuclear
    palsy.
    Neurol 1991; 41: 634-43

55. Pillon B, Dubois B, L'Hermitte F, Agid Y.
    Heterogeneity of intellectual impairment in progressive supranuclear
    palsy, Parkinson's disease and Alzheimer's disease.
    Neurol 1986; 36: 1179-85

56. Portegies P, De Gans J, Lange JMA.
    Declining incidence of AIDS Dementia Complex following introduction
    of zidovudine treatment.
    Br Med J 1989; 299: 819-21

57. Rossor M.
Primary degenerative dementia.
In: Bradley WG, Daroff RB, Fenichel GM, Marsden CD (Eds).
Neurology in clinical practice.
Boston, Butterworth, 1990: 1409-22

58. Squire LR, Zola-Morgan S.
The medial temporal lobe system.
Science 1991; 253; 1380-6

59. Squire LR.
Memory and brain.
New York, Oxford University Press, 1987

60. Strub RL, Black FW.
Organic brain syndromes.
Philadelphia, FA Davis, 1981

61. Subcortical dementia
Cummings JL (Ed).
Oxford, Oxford University Press, 1990

62. Tissot R, Constantinides J, Richard J.
Pick's disease.
In: Frederiks JAM (Ed). Handbook of clinical neurology, vol. 2.
Amsterdam, Elsevier Science Publ, 1985: 233-245

63. Tomlinson BE.
The pathology of dementia.
In: Wells CE (Ed). Dementia, 3rd ed.
Philadelphia, FA Davis Company, 1978: 113-53

64. Van Gool WA.
Alcoholisme en dementie (Alcoholism and dementia.)
Acta Neuropsychiatrica 1991: 3: 26-29

65. Whitehouse PJ.
The concept of cortical and subcortical dementia: another look.
Ann Neurol 1986; 19: 1-6

66. Whitehouse PJ, Price DC, Clark AW et al.
Alzheimer's disease: evidence for selective loss of neurons in the
nucleus basalis.
Ann Neurol 1981; 10: 122-6

CHAPTER 3

Dementia in diseases predominantly affecting subcortical structures

Introduction

In 1974, after studying the neurobehavioural changes in patients with progressive supranuclear palsy (PSP) and reviewing the literature on this subject, Albert et al introduced the concept of *subcortical dementia* [5]. The main characteristics consisted of: forgetfulness, slowing of thought processes, impaired ability to manipulate acquired knowledge, and personality changes, consisting mainly of apathy or depression and irritability. A notable feature was the absence of aphasia, apraxia, and agnosia. The authors proposed that the mechanism underlying subcortical dementia is that of disturbed timing and activation of cognitive processes. Impaired function of the reticular activating systems or disconnection of reticular activating systems from thalamic and subthalamic nuclei was thought to be the cause of slowing down of normal intellectual processes, while cortical systems responsible for perceiving, storing and manipulating knowledge were thought to be intact. They could, however, be abnormally activated; and once activated, they would take an excessive amount of time to carry out intellectual processing [5].

In the following year McHugh and Folstein described the picture of the mental changes in patients with Huntington's disease (HD), and explained these in the framework of the concept of subcortical dementia [190]. Their description is very similar to that of Albert et al. [5]. A common feature in both patient groups was the absence of aphasia, apraxia, and agnosia. In both publications the similarities with other neurological diseases with mainly subcortical pathology were mentioned.

In 1983 Cummings and Benson, in describing dementia associated with various diseases, stressed the difference between cortical and subcortical patterns of dementia [59]. Great emphasis was put on the neurobehavioural characteristics on which this distinction can be made.

Between 1974 and 1983 the number of diseases with possible subcortical dementia was considerably expanded. Included were: Parkinson's disease, Huntington's disease, Progressive supranuclear palsy, Wilson's disease, spinocerebellar degeneration and idiopathic basal ganglia calcification, normal pressure hydrocephalus, toxic and metabolic encephalopathies and the dementia syndrome accompanying depression [46,59]. In the following years subcortical dementia was also described in patients with sarcoidosis, Binswanger's disease, acquired immunodeficiency syndrome (AIDS), and multiple sclerosis [72,74,115,120,151,205,231].

Psychological and behavioural descriptions of subcortical dementias focused on the following characteristics: slowed thinking, impaired ability to manipulate acquired knowledge, deficits in learning new information, and psychopathological symptoms, especially apathy and irritability. A consistent feature is the emphasis on the absence of disturbances of language-dependent activities and other higher cortical impairments (aphasia, apraxia, agnosia) as opposed to the presence of these defects in cortical dementia, with Alzheimer's disease as the prototypical example.

Neurophysiological and anatomic-pathological differences between cortical and subcortical dementia (sub-)types and have been reported [54,99]. According to Whitehouse (1986), discrepancies between the overall clinical and the specific pathologic-anatomical and neurochemical characteristics of cortical and subcortical dementia (sub-)types, challenge the division into only two separate dementia categories [279]. In addition, different approaches in the assessment of cognitive dysfunction have led to conflicting results with regard to the clinical recognizability of cortical versus subcortical dementia syndromes [126,186].

Despite remaining controversies about the validity of the distinction [279], on clinical and neurobehavioural examination there are marked behavioural and neuropsychological differences between patients in the early stages of subcortical and cortical dementia syndromes [19,29,59,62,63,88,125].

When one accepts the clinical distinction between cortical and subcortical dementia, a new problem emerges: if there are several types of subcortical dementia, what are the similarities; is subcortical dementia one clinical entity and can the behavioural and neuropsychological impairments in the different diseases be explained by one underlying mechanism?

These were the central questions, in reviewing the literature on mental changes and dementia in the following diseases:

Progressive supranuclear palsy (PSP)
Huntington's disease (HD)
Parkinson's disease (PD)
Normal pressure hydrocephalus (NPH)
Subcortical arteriosclerotic encephalopathy (SAE), or Binswanger's disease
Acquired immunodeficiency syndrome (AIDS)
Multiple sclerosis (MS)

Emphasis was put upon the literature of the last decade because in these studies neuropsychological tests are used for specification of the types and patterns of functional deficits, often in relation to pathophysiological and neuropathological features.

Before analysing the results of neuropsychological assessments, the clinical manifestations, pathology and pathophysiology of each disease will be described, followed by an elucidation of the criteria used in literature selection.

Clinical manifestations, pathology, and pathophysiology

Progressive Supranuclear Palsy
Progressive supranuclear palsy (PSP) or the Steele-Richardson-Olzewski syndrome, is a non-familial degenerative disease affecting mainly subcortical structures [257,284]. The median age at onset is 65 years [157,177].

The diagnosis is based on the presence of down-gaze abnormalities and at least two or more of the five following features: axial dystonia and rigidity; pseudobulbar palsy; bradykinesia and rigidity; frontal lobe signs; postural instability with falls backwards [163]. Neurobehavioural disturbances and dysarthria are also early features [5].

Studies of brain metabolism with positron emission tomography demonstrated a global decrease in glucose utilization, blood flow, and oxygen utilization but this decrease was most marked in the frontal cortex, which is not pathologically involved in the disease [65,162].

Structures which are severely affected are the globus pallidus, subthalamic nucleus, basal nucleus of Meynert, superior colliculus, periaqueductal grey, substantia nigra, locus coeruleus and dentate nucleus; less affected are the red nucleus, nucleus pontis, and inferior olive; some changes are also found in neostriatum, thalamus and hypothalamus [284].

Huntington's Disease

Huntington's disease (HD) is a genetically transmitted disorder (autosomal dominant trait) of the nervous system resulting in progressive atrophy of the basal ganglia, especially the caudate nucleus. The clinical onset is typically in the fourth of fifth decade [59].

The disorder is marked by a progressive movement disorder (chorea), dementia, and in most cases marked personality changes, e.g. apathy, inertia, depression, increased irritability [59,190,254].

Positron emission tomography studies demonstrated decreased metabolism in the caudate nucleus and putamen, and reduced grey matter blood flow, particularly in the frontotemporal and parietal cortex [21,158,263,288].

Neuropathologically, HD is marked by neuronal loss, and gliosis in the striatum (caudate nucleus and putamen) and, to a somewhat lesser extent in the pallidum; the thalamus may be damaged as well. In later stages, degeneration can also be found in the cerebral cortex, and cerebellum [40,70,84,180,254,274].

Parkinson's Disease

The many causes of parkinsonism are usually divided into three categories: idiopathic, symptomatic, and 'parkinsonism plus' syndromes. The most common form is the idiopathic variety known as Parkinson's disease (PD). This is a degenerative disorder of unknown etiology affecting mainly the pigmented brainstem nuclei. The disease is characterized by release phenomena, e.g. tremor and rigidity, and symptoms due to loss of function, e.g. akinesia and postural abnormality. This produces a complex motor system disturbance including bradykinesia; cogwheel rigidity; resting tremor; masked facies; loss of associated movements; and disturbances of gait, posture, and equilibrium [59,81,255]. Some clinically significant cognitive loss is present in a great proportion of patients with Parkinson's disease [35,37,60,164,201]. But it seems that patients with PD can exhibit one of three patterns of cognitive performance: essentially normal mental status, a small cluster of specific cognitive deficits, or severe and generalized mental impairment [79]. The average reported prevalence of dementia in PD is 25% to 40% [37,60,201,235]. Dementia in PD appears to be more common in those patients in whom bradykinesia rather than tremor predominates [79,123]. The dementia in PD is characterized by cognitive deficits, often personality changes (e.g. depression) and impaired motor function [19,160,173,187,188,201]. The coexistence of dementia of the Alzheimer type accounts for a minority of the dementias of PD [60].

Measurements of cerebral blood flow and metabolism have identified abnormalities in and outside the basal ganglia, e.g. decreased global values in severely affected patients and lower values in frontal regions [216].

Severe neuronal loss occurs in the dopaminergic pars compacta of the substantia nigra and the ventral mesencephalic tegmentum, which project respectively to the striatum and prefrontal cortex. Neuronal loss may be equally severe in the noradrenergic locus coeruleus, serotonergic raphe, and cholinergic basal forebrain [55,59,135].

Normal Pressure Hydrocephalus

Hydrocephalus is a general term denoting an enlargement of the cerebral ventricles. A distinction is being made between noncommunicating (=hydrocephalus secondary to aqueductal stenosis) and communicating hydrocephalus (= ventricular dilation resulting from cerebrospinal fluid (CSF) circulation disturbance in the subarachnoid spaces). Normal pressure hydrocephalus (NPH) is a communicating hydrocephalus with normal CSF pressure, a state originally called occult hydrocephalus [1,59,191,259].

The onset of NPH is insidious, with unresponsiveness and inattentiveness, later followed by the triad of fluctuating but progressive dementia, (urinary) incontinence and gait disturbance [18,149].

In a study with positron emission tomography NPH patients showed a globally diminished cerebral glucose use [129].

The etiology and pathophysiology of communicating hydrocephalus, especially that with normal cerebrospinal fluid pressure, has been actively discussed since 1964/1965 [1,191]. The basic mechanism involves interference of CSF flow. In many cases there is a history of subarachnoid haemorrhage, meningeal inflammation, head trauma, or tumour [18,59]. If no precipitating cause is found the term idiopathic NPH is used.

Pathological findings compatible with NPH are: large ventricles, thinning and even absence of ependyma associated with adjacent periventricular gliosis (demyelination and spongiosis) [75,101,233].

Subcortical Arteriosclerotic Encephalopathy

Subcortical arteriosclerotic encephalopathy (SAE) or Binswangers' disease is an illness of elderly, often hypertensive patients (age group 50-70 year). The disease is considered a form of vascular dementia [10,171,238,242] and is possibly related to the lacunar state [170,238].

The clinical features of SAE are: insidiously progressive dementia; lengthy clinical course with long plateaus and the accumulation of focal neurological symptoms and signs; and often hypertension or systemic vascular disease. Less often a pseudobulbar syndrome with deterioration of gait and sphincter control occurs [52]. The illness may often be interrupted by strokes with partial recovery. Seizures occasionally occur. There are often abrupt clinical changes without typical strokes. The diagnosis currently rests on the clinical features supported by non-specific periventricular white matter changes on CT or MRI; that is, in the absence of other disorders that cause demyelination [10]. Neurobehavioural changes consist of cognitive deficits, depression and apathy [74].

On MRI-spin echo sequences the periventricular white matter shows an increased signal intensity, a radiographic hallmark of demyelination disorders [151]. On CT scanning, the white matter abnormality consists of ill-defined regions of low attenuation, most frequently around the frontal and occipital horns of the lateral ventricles and extending into the centrum semiovale and enlarged lateral ventricles [74,151].

Neuropathologically there is some evidence of deep grey matter ischemia, but the prominent features are: diffuse myelin loss with or without loss of axons, with varying amounts of gliosis and associated lacunar infarcts, and severe thickening of small vessel walls. The leukoencephalopathy is present along the surface of the lateral ventricles and may extend to the cortex, although the corticocortical U-fibres are spared. The boundaries between the lesions and normal white matter are ill-defined [10,118,151,238].

Acquired Immunodeficiency Syndrome

Infection by the human immunodeficiency virus-1 (HIV-1) results in a spectrum of clinical manifestations ranging from asymptomatic seroconversion, or the relatively benign clinical symptoms of the acquired immunodeficiency syndrome (AIDS)-related complex (ARC), to full-blown AIDS [203]. Impairment of the central nervous system may be caused by infection with HIV-1 or by opportunistic infections such as toxoplasmosis and progressive multifocal leukoencephalopathy. Subacute HIV-1 encephalitis evolving into a progressive dementia, the AIDS dementia complex (ADC) is a common neurologic syndrome [68,214,225].

ADC is characterized by cognitive deficits, behavioural disturbances and motor deficits; the progression is stepwise progressive [66,68,72,93,205,225]. ADC is caused by direct infection of the brain with HIV [68,83,107,176,204]. A quantitative comparison of HIV-1 antigen levels in matched serum and CSF specimens indicated that

HIV-1 antigen expression in CSF occurs independently from that in serum and is correlated with ADC [224].

A study with positron emission tomography indicated that hypermetabolism in the basal ganglia and thalamus may be an early component of ADC; as the diseases progresses general hypometabolism develops [240].

Neuropathologically most abnormalities are found in the white matter and subcortical structures, the cortex is relatively spared. The most striking finding is demyelination, focal rarefaction and scattered vacuolization of the central and periventricular white matter and the presence of multinucleated giant cells. They are usually located around small vessels in the centrum semiovale, the basal ganglia and the pons. The cerebral cortex appears to be relatively spared without marked cortical neuronal loss or architectural changes [107,176,204,225].

Multiple Sclerosis

Multiple sclerosis is an inflammatory demyelinating disease of undetermined etiology that involves the white matter of the cerebral hemispheres, brainstem, optic nerves, cerebellum, and spinal cord. It may take the form of an acute, a subacute or a chronic disorder; it may occur with remissions and exacerbations or be slowly progressive. Recurrent episodes of demyelination characteristically produce a relapsing and remitting course [78,182].

Most disabled patients have a chronic progressive (CP) clinical course. This CP pattern can be present from the onset but more frequently it develops after an initial period of relapses and remissions. Sensory and motor disturbances are usually prominent in the clinical presentation and course [78,182]. Dementia is uncommon early in the disease, but cognitive deterioration is frequent in later stages [105,106,217,231]. The true prevalence of cognitive impairment in MS is unknown. Estimates are dependent on the population studied and the method by which impairment is defined. Recent neuropsychological studies have yielded high prevalence rates for cognitive impairment, ranging from 46% in patients with a relapsing/remitting (RR) disease course to 72% for patients who had chronic/progressive (CP) MS [112,232].

Cerebral oxygen utilization and blood flow, measured with

positron emission tomography, have been found to be significantly reduced in patients with MS, with these levels being lowest in patients with cerebral atrophy, and in patients with significant intellectual deterioration [31]. With magnetic resonance imaging virtually all definite MS patients have abnormalities, but there is a considerable variability between the patients [229].

At neuropathological examination the brain grossly appears normal or moderately atrophic. Coronal sections of the cerebral hemispheres reveal a characteristic topographic distribution of the multiple sclerosis plaques, which occur most frequently in the superolateral corners of the lateral ventricles. A smaller number of lesions is distributed throughout the cerebral white matter with a tendency to spare the subcortical U-fibres, cortex, and deep grey structures [59,78,182]. The burden of white matter involvement appears to fall most heavily on frontal white matter [82].

Selection of the literature: criteria

Definition of dementia
A problem in reviewing the literature on cognitive impairment in the aforementioned neurological diseases is the definition of dementia.

The most widely accepted criteria for the diagnosis of dementia are those proposed by the Diagnostic and Statistical Manual of Mental Disorders-III(-Revised). These include [76]: impairment in short-and long-term memory, associated with impairment in abstract thinking, impaired judgement, other disturbances of higher cortical function (e.g. aphasia, apraxia, agnosia, constructional difficulty), or personality change. The disturbance is sufficiently severe to interfere with social or occupational functions, and occurs in the context of clear consciousness. These criteria apply quite well to cortical dementia syndromes, as in Alzheimer's disease, but are often not fulfilled by patients with cognitive deterioration due to subcortical disorders. Especially in the early stages the deficits are demonstrable only on specific neuropsychological tests or experimental tasks, compared with the results of (age-, education-) matched normal controls.

The definition of dementia by Cummings and Benson allows inclusion of both patients with cortical and subcortical dementia syndromes [59]. Operationally, they define dementia as an acquired persistent impairment of neuropsychological function involving at least three of the following spheres of mental activity: language, memory, visuospatial skills, emotion or personality, and cognition. Cognition, in this definition, refers to the ability to manipulate previously acquired information and includes functions like abstraction, judgment, calculation and executive function.

In reviewing the literature both these definitions were encountered. In a substantial number of other publications the presence of global cognitive deterioration and/or dementia is inferred from the results on dementia rating scales or mental status screening tests, without reference to an independent definition.

In the analysis of the neuropsychological findings only those publications have been included in which the presence and/or absence of cognitive deterioration and/or dementia, was clearly defined according to either the DSM-III(-R)[76] or the Cummings and Benson [59] criteria, or based on generally accepted rating scales or screening tests. A further criterium for inclusion was that the test results of patients were compared with severity matched patient groups and/or with matched normal controls, or are expressed as standard scores.

With respect to cognitive impairment and characteristics of subcortical dementia in Parkinson's disease (PD), Huntington's disease (HD), progressive supranuclear palsy (PSP), Acquired immunodeficiency syndrome (AIDS) and Multiple Sclerosis (MS) extensive reviews have been published [28,33,35,62,105,147,163,231]. A review of the literature regarding the memory deficits in HD, PD, MS and normal pressure hydrocephalus (NPH) can be found in Kapur [148]. When necessary these reviews will be referred to. In the current review emphasis is mainly put upon recent publications concerning these subcortical disorders.

Neuropsychological functions and tests
The analysis of the neuropsychological findings concentrated on intelligence, memory, perceptuomotor speed, attention, executive

Table 1. Selection of cognitive functions and examples of neuropsychological tests [166,278]

COGNITIVE FUNCTION	EXAMPLES OF NEUROPSYCHOLOGICAL TESTS
Intelligence	Wechsler Adult Intelligence Scale (Revised) (WAIS(-R)), Raven Matrices
Memory	
– general indication	Wechsler Memory Scale (WMS)
– remote memory	Recall and recognition of famous events, people and scenes
– immediate auditory memory span	Digit Span (forward)
– immediate visual memory span	Corsi block tapping
– auditory verbal acquisition memory, auditory verbal delayed recall and recognition	Logical memory and paired associate learning of the WMS, 10 and 15 words tests, and selective reminding tests
– visual acquisition memory, visual delayed recall and recognition	Visual/visuospatial immediate and delayed reproduction, and visual recognition tests, e.g. visual reproduction of the WMS, Recognition Memory for Faces
– procedural learning	Learning of a motor skill
Perceptual and psychomotor speed	Computerized reaction time tasks, part of the Stroop Colour Word test and Trailmaking test, digit symbol substitution
Attentional activities	Continuous performance tests, digit span (forward and backward), mental control from the WMS, part of the Stroop Colour Word test, and Trailmaking test
Executive control functions (problem solving, set shifting, and abstraction)	Card sorting tests, Stroop interference task, Trailmaking B, Luria frontal lobe series, subtests of the WAIS(-R)
Language	
– naming	Boston naming test
– verbal/word fluency	Controlled oral word association tests
– comprehension	Token test, word-picture matching tasks
– repetition	Repetition of letters, words and sentences
– writing	Assessment of motor characteristics and linguistic quality
– reading comprehension	Comprehension of words, sentences and text
Arithmetic	Calculation tests, WAIS(-R) Arithmetic
Praxis	Performing (symbolic) gestures and activities
Visual information processing (gnosis, perception and visuospatial ability)	Incomplete figure recognition, clock reading, line orientation, Raven Matrices, subtests of the WAIS(-R)
Visuoconstruction	Drawing and copying figures, block design WAIS(-R)
Motor function	Finger tapping, peg board tests

control function, language, arithmetic, praxis, visuoperception, visuoconstruction and motor speed. In *table 1* these functions with examples of specific tests, are presented 166,278.

Neuropsychology and cognitive function

Progressive Supranuclear Palsy
Intelligence: Global intellectual impairment, as measured by performance on the WAIS(R) or the Raven matrices is present in most patients 77,103,130,178,222.

Subtests of the WAIS(R) which show consistent impairment are: similarities, arithmetic, picture arrangement and digit symbol 22,50,103,130,178. The results on the subtests digit span, picture completion, block design and object assembly vary 130,178,222. The subtests information, comprehension and vocabulary are rarely included; if so the results do generally not point to impairment 130,178.

Memory: The total score or Mental Quotient of PSP patients on the WMS indicates memory dysfunction 77,103,167,222.

The results concerning memory for remote and recent events vary, but do not indicate a general impairment 50,130,22.

In two studies auditory immediate memory was defective 50,222. Auditory verbal acquisition memory is impaired 50,130,167,222, but improvement with repetition has been reported 103. Few results regarding auditory verbal delayed recall were found, but these point to impairment, although the rate of forgetting is relatively intact 167. Auditory verbal (delayed) recognition is relatively normal 178,167.

No results regarding visual immediate memory were found. Visual acquisition memory as measured by reproduction of drawings is impaired 222. No results regarding delayed visual recall were found in the literature. Visual recognition memory as measured with recognition of faces is relatively normal 178.

Until now procedural learning has not been examined in PSP.

Perceptual and psychomotor speed: Delayed reaction time, slowed perceptual speed and scanning, and slowed mental or cognitive processing speed is found in patients with PSP 50,77,103,130,167,178. Albert et al described slowing of thought processing as one of the prominent characteristics in the patients included in their analysis [5].

Attentional activities: Attention, concentration and tracking deficits, as measured by neuropsychological tests, are present [50,103,222,227].

Executive control functions: A major characteristic of the mental changes in PSP patients is inability to manipulate acquired knowledge [5].

Results on behavioural rating scales indicate the presence of so-called frontal lobe dysfunction, e.g. imitation behaviour, apathy and loss of initiative, grasping and motor impersistence [50,65,77,95,178,222]. Perseverations occur on neuropsychological tests [50,103,178]. Problem solving and concept formation, measured with sorting tasks, is impaired [77,103,222]. Studies in which (parts of) Luria's frontal lobe tests were employed, report deficits [50,77,222]. Mental flexibility, measured by part III of the Stroop test is reduced [95,103]. Verbal concept formation is defective [50,178]. Verbal reasoning and judgment are impaired [130], as is visual sequential problem solving [50,103,178].

Language: Dysarthria and dysphonia are early symptoms of PSP [5,50,65,163].

Verbal fluency, as measured on tasks of controlled oral word association tests, is reduced. Slight difficulties on naming tasks and/or mild word finding difficulties are sometimes mentioned; comprehension and repetition are normal and features of aphasia are absent [5,50,103,163,178,222].

Handwriting is often normal [50]; features of agraphia have not been reported.

Tests of reading comprehension have not been reported in patients with PSP.

Arithmetic: Performance of basic arithmetical activities is normal in patients with PSP; mildly impaired performance is sometimes found on more complex tasks, when time-limits are applied [50,130,178,222].

Praxis: No apraxia has been reported in patients with PSP [5,50,222].

Visual information processing: Although down gaze abnormalities are among the signs of the disease, basic visual gnosis is normal in patients with PSP [5].

Difficulties and impaired performance have been reported with regard to decoding and analyzing complex visual (-spatial) material [50,62,77,178,222].

Visuoconstruction: Performance of PSP patients on drawing tasks and block construction tasks is (mildly) impaired [50,178,222].

Motor function: Movement disorders are part of the disease. Two studies report (mild) reduced motor speed in PSP patients, for instance as measured on pegboard tests [50,103].

Huntington's Disease

Intelligence: A significant drop in the full scale IQ as measured by the WAIS(R) has been found in all patients with a definite diagnosis, whether they are considered demented or not [21,28,96,144,145,202]. The performance scale seems to account for most of this decline [21,144,145], mainly the following subtests: picture arrangement, digit symbol, object assembly [11,21,144,145]. Sometimes impairment is found on the subtests arithmetic and digit span [144,145]. The results on the subtests information, comprehension, similarities and vocabulary are usually normal, except for mild impairment in more advanced stages of the disease on comprehension and similarities [21,94,144,145]. Performance on the Raven matrices of non-demented patients is lower than that of normal controls [21,275].

Memory: Memory has been extensively studied in HD [28,33,62,148]. In unselected groups of patients, those with advanced HD and in patients with a diagnosis of dementia the MQ of the WMS and comparable global indications of memory function are decreased [21,42,96,143,202,283].

Even early in the course of the disease there are remote memory deficits involving all decades to an equal degree [3,4].

Auditory digit span can be impaired to some (mild) degree, although probably not in recently diagnosed HD [33]. Auditory verbal acquisition memory and delayed recall are defective in patients with (recent and advanced) HD and in patients with a diagnosis of dementia [11,21,33,41,42,44,96,148,184,246,247,283]. With a prolonged exposure to the material to be learned there is a significant improvement [181]. Repetition of the information to be remembered improves performance on repeated trials [41,44].

Results on delayed recall tests are impaired, but the saving scores are relatively intact [41]. Performance on tests measuring auditory verbal (delayed) recognition are (relatively) normal [11,33,41,44,148,247].

No results concerning immediate visual memory function were found. Visual acquisition memory, e.g the reproduction of drawings, is impaired in patients with and without a diagnosis of dementia [11,21,33,44,96,128,144]. The same applies to visual delayed recall, but the rate of forgetting on delayed visual reproduction is relatively normal [11,33,44,202]. Findings with regard to visual, immediate and delayed, recognition memory vary but seem to indicate normal visuoverbal recognition, and defective performance in recognition of nonverbal material in patients with and without a diagnosis of dementia [33,148,181,202,246,247].

Recent interest in procedural learning has shown selective deficits in (non-)demented HD patients [33,41,113,148,181,246].

Perceptual and psychomotor speed: Perceptual speed and scanning, as well as reaction time and psychomotor speed have been found to be delayed in patients with HD [11,44,94,144], but not in subjects at risk for HD [134].

Attentional activities: The performance of patients with a diagnosis of HD on a variety of tasks which require and measure attention, sustained concentration and tracking (e.g. serial sevens, mental control, digit symbol substitution and part of the trailmaking test and stroop test) is mildly to moderately impaired [11,21,28,29,44,62,144,145].

Executive control functions: One of the major behavioural changes noted in HD patients is an apathy, which can be explained as a loss of the ability to initiate activities, a deficiency in planning and sequential arrangement, and in overall arousal [45].

Mild to moderate impairment on complex sequencing tasks has been reported in patients in more advanced stages of the disease [11,28,33,44,144]. Problem solving ability and mental flexibility are compromised, but in general not in early stages of the disease [11,33,96,144,246,277]. Verbal concept formation, verbal reasoning and judgment, and visual sequential problem solving, measured with the subtests similarities, comprehension and picture arrangement of the WAIS are mildly impaired in more advanced stages of HD [43,94,144,145]. HD patients with evidence of dementia are impaired on tasks measuring frontal lobe functions [143,247]; in this subgroup perseverative behaviour can occur [96,143,144,247].

Language: Speech is altered in HD, with a hyperkinetic pattern and loss of normal resonance and prosody [254,275].

Aphasic features are absent [33,45,62]. Performance on object- and picture-naming tasks is relatively normal in different subgroups [11,12,33,42,44,62,143,275]. Verbal fluency is reduced in HD patients with and without mental deterioration [33,42,44,62,143,184,247,283]. Sometimes mildly impaired performance on comprehension tests is reported [11,275], but these do not seem to be aphasic in nature [275].

Handwriting can be impaired, but no linguistic deficits are reported.

Reading comprehension is normal [44].

Arithmetic: Basic arithmetical abilities seem to be intact in HD patients. Sometimes mildly impaired performance has been found on tasks in which attention and cognitive speed are important factors [44,144,145].

Praxis: Although it has been rarely examined, there are no indications for the presence of apraxia in HD patients [33,44].

Visual information processing: Visual agnosia is not a characteristic of HD. Demented HD patients are impaired on visuospatial tasks; findings with regard to complex visuoperceptual ability vary, but mild impairment can be present in advanced HD [11,28,33,45,62,94,144,145,275].

Visuoconstruction: Mild to moderately impaired visuoconstructional ability is frequently reported for HD patients with and without global mental deterioration, but not for patients at risk [28,33,44,45,62,94,143-145,247].

Motor function: A progressive movement disorder is one of the symptoms of the disease. Studies assessing neuropsychological aspects of motor function in HD patients were scarce, but results indicate slowing of motor speed in patients with HD, but not in patients at risk [44,94,144].

Parkinson's Disease

Intelligence: Results on intelligence measures are decreased in patients with a diagnosis of dementia [23,222].

Performance on the verbal scale of the WAIS seems largely unaffected and the main deficits concern subtests of the performance scale, e.g. block design, picture arrangement and digit symbol [23,116,121,122,160,235,261].

Findings concerning the Raven matrices vary, but often impaired performance is found [49,121,122,221,222].

Memory: Memory disturbances are one of the characteristics of the dementia in PD [33,35,62,148]. General indications of memory function, e.g. MQ, are lowered in patients with and without a diagnosis of dementia [23,77,79,111,160,220,221,261].

Remote memory impairment, with equal involvement of all decades, was found in patients with evidence of intellectual deterioration [33,35,122,125,148,244], but not (generally) in non-demented patients [26,125,244].

Reported results on tasks like (auditory) digit span forward, are normal in the different patient groups [26,33,73,116,117,121,122,125,241,264]. On several tests deficits in auditory verbal acquisition memory are frequently found in PD, most often in patients with global cognitive deterioration and/or a diagnosis of dementia [23,33,35,49,62,73,79,114,121,122,125,126,148,160,221,235,243,244,261]. The results of patients without symptoms of dementia or cognitive deterioration vary sometimes but are mostly within the normal range [49,125,243,244,261]. When rehearsal is provided learning improves and PD patients benefit from recall probes [79,114]. PD patients with evidence of dementia are impaired on auditory verbal delayed recall; the results of nondemented patients vary but are usually within the normal range [23,33,35,49,73,79,114,148,183,261,264,265]. The rate of forgetting from immediate to delayed recall is normal, even in patients with evidence of dementia [79]. Performance on tasks measuring auditory verbal recognition memory is normal [33,35,79,114,148,265].

In the few studies in which immediate visual memory span was measured no significant abnormalities were found [200,261]. Visual acquisition memory and delayed recall, examined with a variety of paradigms and test materials, seem only to be impaired in patients with mental deterioration [23,33,35,49,260,261,264,265]. No significant deficits on visual recognition memory tests have been reported [33,35,73,261,264,265].

Procedural learning, like acquisition of a motor skill (although few studies are available) seems to be impaired PD, especially in patients with dementia and in more advanced stages of the disease [111,113,246].

Perceptual and psychomotor speed: Significantly slowed performance on tasks measuring reaction time and psychomotor speed is found in PD patients with and without mental deterioration, and with early and late onset of the disease [24,26,73,94,116,117,125, 183,200,220,237,245,264,265].

Attentional activities: Parkinsonian patients with evidence of mental deterioration and/or a diagnosis of dementia, manifest abnormalities on tests of attention and especially concentration and tracking [23,33,62,73,160,235]. Sometimes with experimental tasks mild impaired performance is also found in PD patients without mental deterioration [34]; but this is not a general finding when neuropsychological tests are used for the different aspects of attention and concentration [97,100,160,165,183,264].

Executive control functions: Between 1975 and 1985 a considerable amount of research has focused on and reported 'frontal-type' dysfunction in PD [33,35,201].

A defective planning or a defect in the regulation of behaviour seems to play an important part in the cognitive impairment in demented PD patients [73]. Results on behavioural rating scales indicate the presence of so-called frontal lobe dysfunction, e.g. imitation behaviour, apathy and loss of initiative, grasping and motor impersistence, even in patients without evidence of mental deterioration [77,218,221,222,250]. Mild reduced mental flexibility and problems of complex sequencing have been reported [116,117,265]. Visual concept formation and complex problem solving are impaired and perseverations can occur in different patient subgroups [23,77,87,88,100,165,200,221,222,246,250,256, 264,265].

Verbal reasoning and judgement does not seem to be defective [36,100,200,235]. Verbal concept formation can be impaired but only in patients with evidence of dementia or patients with severe symptoms of the disease [117,122,160,221,235]. Results regarding visual sequential problem solving indicate deficits only in patients with more global cognitive impairment [36,261].

Language: Speech of PD patients is characterized by hypophonia, dysarthria and a monotonous quality [59,60,73].

Incidentally mild impairment on naming tasks is reported in patients with evidence of dementia but in general no significant deficits have been found [12,33,60,62,73,79,122,222,235]. (Mildly) reduced verbal fluency is a general finding in demented patients [33,49,60,62,73,122,222]. Slight comprehension difficulties are sometimes reported in demented patients and also problems with judgement [60,121,122] but this is not a general finding [33,62,73,114,121,122,125,235]. Repetition is usually normal [60,73,234].

Defective handwriting, especially micrographia is a well-known symptom, but the linguistic quality of writing is normal [59,60,73].

Although reading has been rarely examined, significant impairment was not reported [60].

Arithmetic: Arithmetic has seldom been examined in patients with PD. Patients with evidence of mental deterioration are sometimes impaired on the arithmetic subtest of the WAIS(R) or on 'serial sevens', but this is not a general finding [73,122,125,235]; if present, impaired performance on these tasks is probably caused by time-constraints or attentional difficulties [73].

Praxis: In general, apraxia is absent in patients with PD and mental deterioration [33,121,122,125,160,222].

Visual information processing: Basic visual gnosis is normal in PD. Performance on complex visuoperceptual and especially visuo-spatial tasks is impaired in PD patients with evidence of dementia [23,49,73,114,121,122,125,160,221,235] and sometimes in unselected patient groups and in patients with normal intellectual function [33,35,62,77,97,116,125,128,160,165,198,220,221,228].

Visuoconstruction: Drawing and constructional ability are impaired in PD patients with evidence of dementia, and this is also reported in most of the studies of unselected patient groups and of patients without evidence of mental deterioration [23,33,35,49,62,73,94,111,116,117,121,122,235].

Motor function: A complex motor system disturbance is present in PD. Although no findings were found with regard to patients with mental deterioration, speed of performing motor tasks is slow in patients with PD [34,36,116,117,264].

Normal Pressure Hydrocephalus

Intelligence: Literature regarding psychometric and neuropsychological test performance of patients with NPH is scarce, in spite of numerous clinical descriptions. In older studies a considerable decrease of the WAIS IQ is mentioned, with some improvement after shunting; seldom the patients reached their estimated premorbid level [1,56,248,249]. After shunting, most improvement was shown on the subtests similarities and comprehension; no improvement was shown on the subtest digit symbol [248]. Performance on the subtest vocabulary is not significantly different from normal controls [110].

Memory: There are only a few studies in which memory functions in patients with NPH are thoroughly examined.

In clinical descriptions memory loss is frequently mentioned [1,30,53,212,262]. In some studies memory deficits were only mild [148,281], in others a significant proportion of the patients demonstrated a general memory loss [56]. In some studies performance of patients before and after shunting is compared, showing improvement, but the results of the preoperative assessment are in general not compared with controls, nor expressed in standard scores [101,102,110,266,281].

No results of remote memory testing were found in the literature, the same implies to immediate auditory and visual memory span.

Auditory verbal acquisition memory impairment has been reported [1,53,110,212], as well as defective auditory verbal delayed recall [53,273]. No information was found on the possible effect of repetition or the use of probes on recall performance. Auditory verbal (delayed) recognition seems to be intact [273].

Visual acquisition memory is impaired [53,110]. No reports were found on visual delayed recall, visual recognition memory and procedural learning.

Perceptual and psychomotor speed: In descriptions of the mental changes or behavioural characteristics of patients with NPH, psychomotor and mental slowing is a prominent feature [1,18,59,110,149,191,212].

Reduced psychomotor speed has been found with neuropsychological tests [273], and on reaction time tasks with some improvement after shunting [266,281].

Attentional activities: (Mildly) impaired vigilance and attention is mentioned in descriptions of the neurobehavioural characteristics of NPH [18,53,59,212,258,262]. Preoperative performance on tasks measuring aspects of attention and concentration was reduced in studies examining the effect of shunting [266,281].

Executive control functions: The most common behavioural changes seen are: apathy, euphoria, irritability and disinhibited social behaviour [1,18,59,110,258]. Impaired abstraction, verbal reasoning and judgment, and poor performance on tasks requiring sequential analysis are frequent mental changes in NPH patients [18,59,258]. The presence of frontal lobe signs has been described [1,56]. Occurrence of perseverations on neuropsychological tests has been reported for 7 out of a group of 12 patients [273].

Language: Reviewing the literature, no studies were found in which the different aspects of speech and language function are examined in patients with NPH.

In a clinical description of patients with NPH no dysarthria was found; reduced initiation of conversation was a characteristic finding; the more advanced cases spoke quietly and slowly, sometimes just whispering [212]. Although specific test results are not reported, it has been mentioned that features of aphasia are absent, but comprehension was sometimes slow and laborious [212,262,273]. Slow initiation of speech is sometimes reported in clinical descriptions of the symptoms of NPH [59,212,258,262] and sometimes parkinsonian signs have been reported [149,262]. Reading comprehension seems normal [56].

Arithmetic: No neuropsychological studies were found. In a clinical description of cases, slow and often inaccurate calculation is mentioned [212].

Praxis: Based on clinical descriptions, apraxia seems to be absent in patients with NPH [18,56,59,262,273], with the exception of the so-called 'gait-apraxia' [258].

Visual information processing: Impaired visual gnosis has not been mentioned in clinical descriptions [18,56,59,262,273]. Relying on these few clinical descriptions being available, impaired visuoperceptual and/or visuospatial abilities are not prominent in this disease, although the impaired performance on visuoconstructional tasks seemed to indicate related difficulties [110].

Visuoconstruction: No clear-cut information was found. The results of a few studies regarding the effects of shunting point to impaired preoperative performance on drawing and block construction, with slight postoperative improvement [53,110,266].

Motor function: Part of the triad of symptoms in NPH is a gait disturbance. No information was found with regard to performance on neuropsychological tests of motor function. Clinical descriptions do mention slowing of physical processes [18,59,110,212,262].

Subcortical Arteriosclerotic Encephalopathy

Intelligence: In most reports concerning cognitive function of patients with SAE general intellectual impairment is mentioned, but this is often vaguely described [10].

In one study impairment on the verbal scale and in two studies impairment on the performance scale of the WAIS is described [161,172]. Analysis of the performance subtests showed most impairment on picture completion, block design, picture arrangement and digit symbol [161].

Memory: Memory impairment is frequently mentioned in clinical descriptions of patients with SAE [10,52,131,151,238,242], but there were only a few studies in which neuropsychological performance of patients has been analyzed. Measured with the WMS the MQ is significantly lowered [161].

No significant impairment is found with regard to remembering recent and remote events and personal information [74,161].

The results on auditory digit span tests are normal [74,161]. SAE patients are impaired on auditory verbal acquisition memory tasks [74,161]. Neuropsychological and/or psychometric results of auditory verbal delayed recall and recognition are lacking. The aforementioned clinical descriptions point to impaired performance on delayed recall tasks and intact recognition [51]; based on our clinical experience, repetition of the information improves performance on immediate and delayed recall [74] (Derix, unpublished observations).

Visual acquisition memory is impaired [74,161]. Results with regard to visual delayed recall, visual recognition memory and procedural learning were lacking.

Perceptual and psychomotor speed: Prolonged latency in response to questions or requests is one of the features in descriptions of the mental deterioration in SAE [51,52]. Slow and/or fluctuating speed of responding on different tasks has been found in patients with SAE [74]. In a sample of 41 elderly patients with a history of minor stroke and leukoaraiosis on MRI, of whom 88% were hypertensive, performance on tasks measuring speed of information processing was impaired and related to leukoaraiosis [146], although the diagnosis of SAE was not made.

Attentional activities: In clinical descriptions of patients with SAE attention and/or concentration deficits were frequently mentioned [10,51,52].

Performance on neuropsychological tests has been found to be impaired [74,161,242]. In the aforementioned sample of 41 patients in which slowed speed of information processing was found, performance on tasks measuring attentional activities was impaired and related to leukoaraiosis [146].

Executive control functions: One of the clinical characteristics in SAE is the development of an abulic state, characterized by decreased spontaneity and quantity of behavioural activities, prolonged latency in responding, and difficulty persevering with a task [51,52]. This abulic state was interpreted as a reflection of frontal lobe involvement [51].

In a clinical description of 15 patients, frontal lobe dysfunction and poor conceptual and reasoning ability was reported [172]. In a review of pathologically verified cases published between 1912 and 1985, marked difficulties of judgement were thought to be present in later stages of the disease [10]. Low performance on tasks measuring abstract and concrete thinking has been mentioned, but no test results were given [242]. Severely impaired performance has been reported on the subtest picture arrangement of the WAIS, and was interpreted as an indication of poor social reasoning and sequential thinking [161].

Language: Speech of SAE patients is usually abnormal with dysarthria and dysphonia, and impairment of rhythm and prosody [10,51,52,74,131,172].

In clinical descriptions of patients with SAE, features of language impairment are rarely mentioned [10,151]; difficulties in reading and writing are sometimes described in clinical case presentations but the clinical details were insufficient to determine the underlying cause [51]. Except for a reduced verbal word fluency, no specific language problems have been found in two recent studies [74,242]. Lee et al (1989) reported minimal receptive impairment, measured by performance on the Token test [161]. In the study by Derix et al (1987) only two patients showed aphasic deficits, but these could be explained by the presence of cortical infarcts [74]. In this study only one patient demonstrated difficulties on writing to dictation, which consisted of perseverative errors; one patient showed a tendency to micrographia. Reading comprehension was unimpaired [74].

Arithmetic: Arithmetic, at least with regard to basic ability, is intact [74]. Acalculia has not been mentioned in clinical descriptions of patients with SAE [10,51,52].

Praxis: Apraxia has not been found in patients with SAE [74] and is not mentioned in clinical descriptions of patients [10,51,52].

Visual information processing: The few data available did not indicate difficulties on basic visual gnostic tasks [10,51,52,131,242]. Performance on complex visuoperceptual and visuospatial tests is impaired [74,161].

Visuoconstruction: The scarce information available indicates impaired performance on tasks like drawing and block construction [74,131,161,242].

Motor function: SAE can be accompanied by focal motor impairment and/or deterioration of gait. With regard tot neuropsychological tests there was no information available. In clinical descriptions of SAE patients, (sub)acute motor deficits have been reported [10,51,52,151].

60 |

Acquired Immunodeficiency Syndrome

Intelligence: In most studies of intellectual functioning in patients with AIDS, the evidence for the presence of dementia was inferred from the presence of intellectual and cognitive deterioration. Rarely, a clinical diagnosis of Aids dementia complex (ADC) was given beforehand. In one study in which patients with a clinical diagnosis of ADC were studied, general intellectual deterioration was not always present [72].

A significantly lowered full scale IQ and verbal IQ on the WAIS-R in patients with AIDS was found in two studies [241,272]. One study reported a decrease of performance IQ [272].

No specific profile could be found in patients with AIDS with regard to the results on the different subtests of the WAIS(R) no specific profile could be found in patients with AIDS; the results on the individual verbal and performance subtests are usually within the normal range [39,104,241,272]. Patients with AIDS and/or a diagnosis of ADC often show defective performance on the digit symbol subtest [268,286,287].

Memory: Memory impairment has often been mentioned in clinical descriptions of demented patients with AIDS [205]. No results were found in the literature with regard to measures like the MQ of the WMS. In patients with a clinical diagnosis of ADC a global memory impairment is not always present [72].

Until now, no results have been published regarding memory for recent and remote events in patients with AIDS, with or without mental deterioration.

In different groups of patients with ADC, AIDS and ARC, with and without cognitive impairment, results on tests measuring auditory verbal immediate memory were within the normal range [132,189,197,214,268,272]. Patients with ARC and a diagnosis of dementia, demented patients with AIDS and patients with a clinical diagnosis of ADC were impaired on tests for auditory verbal acquisition memory [72,251,286,287]. Results in AIDS and ARC patients without a diagnosis of dementia or evidence of cognitive deterioration varied. Overall, they indicate that these memory deficits are often present, but are relatively mild [39,72,104,197,251,268,272]; with repetition, there is an increase in the information recalled [214]. (Mild) defective auditory verbal delayed recall has been found in ARC patients with cognitive deterioration [104,132,214,251] and sometimes in patient with AIDS [104,214,251], but this was not always the case [39,272]. No results were reported for patients with a clinical diagnosis of ADC, although analysis of the data in one study demonstrates clear deficits. In these patients recognition memory was relatively intact [72]. In the few studies examining auditory verbal recognition memory, no significant impairment was reported [189,272].

Results on visual immediate memory tasks were lacking. Visual acquisition memory is impaired in patients with ADC [240]. One study demonstrated the same deficit in an unselected population of patients with AIDS, but 25% of these patients proved later to be dementing [39]. In other patient groups no significant impairment has been found [104,214,268,272]. In one study impaired visual delayed recall in AIDS patients is reported [272]. Results of patients with ADC were lacking, and findings reported in other groups were few and varied between no impairment and mild impairment [104,214,268]. No data were found on visual recognition memory and procedural learning.

Perceptual and psychomotor speed: Patients with AIDS and with ARC, and especially patients with the clinical characteristics of ADC or evidence of mental deterioration show prolonged reaction times, and slowed psychomotor and perceptual speed [72,133,147,189,215,240,241,251,268,272,286,287].

Attentional activities: Patients in the initial stage of ADC frequently complain of difficulties with attention and concentration; corresponding deficits were found on clinical evaluation [120,203,205].

On neuropsychological examination patients with AIDS and with ARC, and especially patients with a diagnosis of ADC or mental deterioration are impaired on the majority of tests used for assessing attention, concentration and tracking [72,104,197,214,240,241,251,268,272,286,287].

Executive control functions: Common behavioural disturbances in patients with ADC are: apathy, loss of spontaneity, and social withdrawal [203,205]. In descriptions of the clinical aspects of AIDS, one of the neuropsychological features was impaired sequential-alternation problem solving and complex sequencing [225,226].

(Mild) difficulties on complex sequencing tasks have been reported in patients with AIDS and evidence of mental deterioration and in patients with a clinical diagnosis of ADC [189,240,268], but not always in other subgroups [39,104,214,241,251,272]. Mildly impaired mental flexibility, as measured on sorting tasks and part III of the Stroop test has not been a consistent finding in patients with AIDS or ARC [104,214,241]. This cognitive function has not been examined in patients with a clinical diagnosis of ADC. Verbal concept formation, verbal reasoning and judgment, and visual sequential problem solving, measured with subtest of the WAIS-R, are relatively normal in patients with AIDS and patients with relatively mild global cognitive impairment [156,214,268,272].

Language: Dysarthria and slowing of speech have only been incidentally described in patients with ADC [203,205,225].

In general, performance on naming tasks is normal in patients with and without evidence of dementia [72,240,268,272]. Mildly reduced verbal fluency is sometimes reported in patients with evidence of mental deterioration or patients in an advanced stage of the disease only [72,156,189,214,240,268,272]. No study reported impairment of receptive language, repetition, judgment of similarities, and reading.

Deterioration of handwriting is sometimes described [205]. As far as examined, linguistic quality of writing was normal [72]. No study reported difficulties with reading.

Language impairment, if present, could be explained by focal lesions caused by cerebral opportunistic infections (mostly toxoplasmosis) [72,206].

Arithmetic: Impairment of calculation has not been reported for patients with AIDS and/or ADC [72,203,205].

Praxis: Apraxia as a deficit of higher cortical function is, in general, not present in patients with AIDS and/or ADC; an opportunistic cerebral infection was the most probable cause when this disorder was found [72,203,205].

Visual information processing: Based on the information available with regard to performance on the picture completion subtest of the WAIS-R [39,272], and based on our experience in examining patients with AIDS with and without ADC very mild difficulties on complex visuoperceptual and/or visuospatial tasks are sometimes present. These difficulties most often seemed a sequela of abnormalities in the visual system (Derix, unpublished personal observations).

Visuoconstruction: Assessment of drawing was lacking in the literature. From our own experience some difficulties in performing drawing tasks can be present in patients with ADC (Derix, unpublished personal observations). Performance on the block design subtest of the WAIS-R is mildly impaired in patients with ADC [189,240,268]. In unselected patient groups, in AIDS patients with normal intellectual functioning and in patients with ARC performance on this test, was mostly normal [39,156,214,268,272].

Motor function: One of the clinical symptoms of ADC is the presence of motor deficits. The results of studies in which neuropsychological aspects of motor function were examined indicate reduced speed in different subgroups of AIDS patients [189,197,240,268,286,287].

Multiple Sclerosis

Intelligence: In most studies criteria for diagnosing the presence of dementia or global cognitive impairment were not specified. In studies using appropriate criteria a lowered full scale IQ on the WAIS(R) was found [192,269], or can be inferred from the results [82]. Some studies reported lowered verbal IQ [192,269], and a lowered performance IQ in MS patients with dementia [82,269]. Studies assessing intellectual functioning in patients with cognitive deficits did not demonstrate general deterioration, although longitudinal studies seem to suggest that intellectual functions deteriorate slightly over time [231].

Findings concerning the individual subtests of the WAIS(R) varied. Decreased performance, if present, was sometimes found on block design [82,112,175,269], object assembly and digit symbol [82,112,269], similarities [112,175], picture arrangement and arithmetic [112,269]. Patients with dementia were impaired on the subtest comprehension [269]. Although performance on the digit span subtest seems consistently lower, relative to the other verbal subtests [231], the scores are within the normal range in patients with and without a diagnosis of dementia [82,112,168,269].

In one study of patients with MS and cerebral white matter abnormalities on MRI scanning, lowered performance on Raven Matrices was found [9]. Normal results were found in a longitudinal study of an unselected patient group [136,139].

Memory: Memory deficits have often been observed in patients with MS and have been extensively investigated, although evidence for the presence of dementia and/or cognitive deterioration was often lacking in most of these studies. These studies generally found that MS patients are impaired in their ability to learn and recall information compared with normal controls, although considerable variability was observed within groups of unselected patients [105,217,231]. In recent studies, the MQ of unselected MS patients was not significantly lowered [136,138,168,175].

As far as examined, remote memory deficits were found only in chronic progressive MS patients [15,16].

Short term memory or auditory verbal immediate memory was unimpaired in unselected patient groups [16,136,148,168,175,229,230,231, 269], and in patients with relapsing-remitting MS [15,112]. In only two recent studies deficits in patients with chronic progressive MS were reported [15,124], in three others these deficits were not found, although in one of those the patients were considered demented [27,82,112]. Auditory verbal acquisition memory has been investigated with a variety of tests and most studies reported deficits in patients with mental deterioration and/or chronic progressive MS, and sometimes lowered performance in unselected patient groups and in patients with relapsing-remitting MS [15,16,44,85,112,124,138,139,148,168,175,192,217,231,269,271]. Studies on auditory verbal delayed recall showed the same pattern, although impairment was found more often [15,16,44,112,136,148, 168,192,231,271]. With repetition the performance improves, with the learning curve slope sometimes comparable to normal controls [15,16,168,169,232,271]. Although results on immediate and delayed recall are lowered, the rate of forgetting is often not significantly different from normal controls [168,232]. Results of auditory verbal recognition tests are relatively intact or even normal in the different patient groups [15,16,44,47,138,139,231,269,271].

A few studies examined immediate visual memory span, with normal results [9,138,139]. Mild to moderate visual acquisition memory deficits, on different tests, have been frequently reported in all subgroups of patients [9,15,16,44,82,85,112,138,139, 175,231,269]. Significantvisual delayed recall impairment was only reported for patients with mental deterioration [85]. Visual recognition memory seems relatively intact [138,139,231,269].

No results were found regarding procedural learning.

Perceptual and psychomotor speed: In MS patients and specifically in patients with chronic progressive MS, with and without global mental deterioration, reaction times are prolonged and psychomotor and cognitive speed are slow [15,44,112,136,137,168,231,271].

Attentional activities: Until 1986 little information was available on attention, concentration, and tracking ability in patients with MS [231]. Recent studies report impaired performance on a variety of tests. The impairment was most prominent in patients with evidence of global mental deterioration, followed by chronic progressive MS. However, mild impaired performance on relevant tests has been reported in patients with relapsing remitting MS and in unselected patient groups [44,47,82,85,112,124, 192,269,271].

Executive control functions: From a review of studies that have used one or more measures of concept formation, e.g. abstract-conceptual reasoning and learning, it can be concluded that MS patients have difficulties forming concepts and shifting sets and sometimes show a tendency to perseverate [231].

Mild impairment on complex sequencing tasks has been reported, especially in chronic progressive MS patients [82,112,271]. Verbal reasoning and judgment and concept formation are normal, although some studies reported very mild impaired performance in a subgroup of MS patients with evidence of cognitive impairment [82,112,168,269]. Mild impaired visual sequential problem solving was found in (chronic-progressive) MS patients with cognitive impairment [82,112,269]. Performance on the category sorting tests measuring problem solving, concept formation and set shifting, can be mildly impaired in patients with chronic progressive MS, MS patients with evidence of cognitive deterioration and MS patients with evidence of (extensive) cerebral involvement [9,13,14,47,82,112,229].

Language: Articulatory disturbances are often found in patients with MS [59,78,182]. Linguistic functions have not been extensively studied in MS patients [231].

Recent studies did not report marked impairment on naming tasks, although very mild difficulties were sometimes present in patients with evidence of global mental deterioration [15,44,124,136, 137,175,229,231,269]. Verbal word fluency is (mildly) reduced in different subgroups of patients with MS [15,44,82,112,124,136,229,231,271]. All other language functions, including linguistic quality of writing and reading comprehension are unimpaired [9,44,82,124, 136,137,269].

Deterioration of handwriting has been reported [44,136,137].

Arithmetic: On simple calculation tasks no deficits have been found [144,175]. Mildly impaired performance has been found on the arithmetic subtest of the WAIS(R) [112], but this was not a general finding, not even in patients with symptoms and signs of global mental deterioration [82,168,269].

Praxis: Praxis has rarely been studied in patients with MS, but apraxia was not reported in recent studies of patients with and without intellectual deterioration [44,124,231,269].

Visual information processing: Although neuroophthalmic abnormalities are among the earliest signs of MS [78,182], clinical cases of MS patients with focal impairment of visual information processing have not been reported [231].

A variety of tasks has been used to assess visuoperceptual and visuospatial ability in different subgroups of patients. Mild impaired performance if present, was mostly restricted to patients with chronic progressive MS and/or patients with evidence of cerebral lesions with evidence of mental deterioration [85,112,269].

Visuoconstruction: Visuoconstructional ability has been examined with drawing tasks and cube construction. In general no deficits were mentioned; findings of (mild) impairment were restricted to patients with cerebral lesions, and patients with chronic progressive MS [44,82,112,136,137,142,175,269].

Motor function: Motor disturbances are usually prominent in the clinical presentation and course of multiple sclerosis [59,78,182].

Motor speed, as measured with fingertapping and pegboard tests, is slowed in the different subgroups of MS patients [44,82,112,140,141,168,271].

Comparison between different subcortical dementia syndromes

There are only a few studies in which a comparison was made between the neuropsychological test results of the various patient groups described in this chapter.

One study compared cognitive deficits between patients in the early stages of HD, and patients with relapsing remitting MS. The groups were matched for age, education and functional performance. The overall impairment of both groups was similar, although the memory deficits of HD patients were more severe and they also experienced difficulties in performing arithmetical computations [44].

A study comparing procedural learning in PD and HD patients reported some differences between patients in an early stage of their disease despite similar motor functioning and unimpaired intellectual functioning. While in early PD patients procedural learning was always impaired, this was not always the case in patients with early HD. Impaired procedural learning was always present in patients with advanced HD, with evidence of some intellectual deterioration, but they were not compared with matched PD patients [246].

In a study examining motor learning and lexical priming in demented and nondemented PD and demented HD patients (compared with age-matched controls), HD patients were impaired on the motor learning task, but not on the lexical priming task. Demented PD patients, but not the nondemented, were impaired on both tasks [113]. A problem in interpreting the results of this study is that, although matched on a dementia rating scale, the demented HD patients were considerably younger than the PD patients.

Differences have been reported between PSP and PD patients with regard to cognitive processing speed, measured with experimental tasks. Both groups were only mildly impaired on intellectual and memory measures, when compared with controls. PSP patients required longer processing time for more complex situations than either normal controls or PD patients and this deficit was associated with impairment in frontal lobe test performance. Frontal lobe impairment was present in both groups but was more severe in PSP patients. PSP patients were also more impaired on tasks for clinical assessment of cognitive slowing [77].

One study compared the patterns of cognitive and behavioural impairment in patients with PSP, with PD, and with dementia of the Alzheimer type (DAT), and normal controls. The groups were matched for age and educational level. Although there were no significant group differences for the degree of intellectual deterioration, measured with verbal tests and a visuospatial tests, the scores of patients with PD fell between those of patients with DAT and PSP. PSP patients performed slightly worse than PD patients on frontal lobe tests and more frontal-type behaviour was present. Both patients with PD and PSP could be distinguished from DAT patients on the basis of frontal dysfunction and, although impaired, better verbal memory function [222].

Subcortical dementia and cognitive function: a summary
The results of the literature review of subcortical dementia and cognitive function are summarized in *table 2*.

There are many reasons why performance on a task of cognitive function can be impaired. Labelling a deficit, e.g. intellectual deterioration, is not the same as explaining the nature of the underlying processes. Clinical neuropsychological tests, as were used in most of the studies reviewed, frequently have the disadvantage of being conceptually complex, which implies that several factors may account for failure in performing them.

Intelligence
Performance on tests like the WAIS(-R) is lowered in patients with evidence of mental deterioration.

Subtests of the WAIS(R) which are most affected are similarities, picture arrangement, and digit symbol. Decreased performance is sometimes reported for the subtests compre-

Table 2. Characteristics of cognitive function in subcortical dementia
syndromes

	PSP	HD	PD	NPH	SAE	ADC	MS
Intelligence	+	+	+	+	+	+	+
Memory							
– general indication	+	+	+	+	+	(+)	(+)
– remote memory	+	+	+	?	+	?	+
– immediate auditory memory span	+	+	-	?	-	-	-
– auditory verbal acquisition memory	+	+	+	+	+	+	+
– auditory delayed recall	+	+	+	+	(+)	+	+
– auditory delayed recognition	-	-	-	-	(-)	-	-
– immediate visual memory span	?	?	-	?	?	?	-
– visual acquisition memory	+	+	+	+	+	+	+
– visual delayed recall	?	+	+	?	?	+	+
– visual (delayed) recognition	-	+	-	?	?	?	-
– procedural learning	?	+	+	?	?	?	?
Reaction time, perceptual and psychomotor speed	+	+	+	+	+	+	+
Attention and concentration	+	+	+	+	+	+	+
Executive control functions							
– complex sequencing	?	+	+	+	(+)	+	+
– mental flexibility	+	+	+	?	(+)	+	+
– visual concept formation	+	+	+	?	+	+	+
– verbal concept formation	+	+	+	+	+	-	+
– verbal reasoning and judgement	+	+	(-)	+	+	-	+
– visual sequential problem solving	+	+	+	+	+	-	+
– 'frontal lobe function'	+	+	+	+	+	?	?
– perseveration	+	+	+	+	?	?	?
Language							
– speech	+	+	+	+	+	+	+
– naming	-	-	-	(-)	-	-	-
– verbal fluency	+	+	+	(-)	+	+	+
– comprehension	-	+	+	+	+	-	-
– repetition	-	?	-	?	?	-	-
– handwriting	-	+	+	?	?	+	(+)
– linguistic aspect of writing	-	-	-	?	-	-	-
– reading comprehension	?	-	-	-	-	(-)	-
Arithmetic							
– basic operations	-	-	-	-	-	-	-
– complex computations	+	+	+	+	(+)	-	+
Praxis	-	-	-	-	-	-	-
Visual information processing							
– visual gnosis	-	-	-	-	-	-	-
– perception	+	+	+	(+)	+	+	+
– visuospatial ability	+	+	+	(+)	+	+	+
Visuoconstruction							
– drawing	+	+	+	+	+	+	+
– block construction	+	+	+	+	+	+	+
Motor speed	+	+	+	+	?	+	+

+ = impairment, - = no impairment, () = little information available, ? = no information available
PSP = progressive supranuclear palsy, HD = Huntington's disease, PD = Parkinson's disease,
NPH = Normal pressure hydrocephalus, SAE = Subcortical arteriosclerotic encephalopathy,
ADC = Aids dementia complex, MS = Multiple sclerosis

hension, block design and arithmetic. Similarities, comprehension and picture arrangement require intact concept formation and reasoning ability. Digit symbol is a test of psychomotor performance. Block design is a measure of visuospatial organization and employs a time-constraint. Performance on arithmetic is dependent on basic arithmetical ability but also requires normal immediate memory (span), concentration and ability to manipulate previously acquired knowledge; time limits influence the total score [166]. IQ based on the performance on Raven Matrices is lowered. This test is widely used as a test for general intellectual ability, but is particularly sensitive to reasoning and visual information processing disorders [166].

Taken together, the mild to moderate intellectual deterioration is caused by impaired performance on tasks requiring attention, concept formation, reasoning ability, cognitive speed, and visual information processing capacity.

Memory
Memory impairment is a general finding although some functions seem to be (relatively) spared. This concerns auditory delayed recognition and probably visual delayed recognition.

Immediate auditory memory span can be slightly reduced in PSP and HD. Immediate visual memory span is normal in PD and MS; no information was available with regard to PSP, HD, NPH, SAE and ADC.

Procedural learning is defective in PD and HD; this function has not been examined in the other diseases.

Studies on auditory acquisition memory report impairment on immediate recall but patients do benefit from repetition, as can be inferred from the acquisition rates on word-list learning tests. Results on verbal delayed recall tests are low, but the rate of forgetting is often comparable to normal controls. The fact that recognition memory is not impaired, indicates that the main memory deficits in subcortical dementia consist of: slowed and less efficient processing and encoding of new information, and impairment of the ability to utilize information stored in the memory system.

Perceptual and psychomotor speed
Without exception perceptual scanning speed, reaction time and psychomotor speed are slow. This is independent of the degree of motor involvement. However, one has still to be careful in interpreting this as a phenomenon of general non-specific cognitive slowing.

In a recent study PD patients were compared with controls on a task measuring higher level cognitive function of planning - a computerized adaptation of the Tower of Hanoi puzzle. PD patients needed more time to make the first move but the number of moves required for solving the problem, and the time for executing subsequent moves was not different from the controls [200]. This seems to imply that cognitive slowing is not a general finding and slowed performance may be task-specific.

Attentional activities
Attentional activities include attention, concentration and conceptual tracking. Attention refers to the capacity for selective perception. Concentration is an effortful, usually deliberate, and heightened state of attention in which irrelevant stimuli are selectively inhibited. Tracking involves attentively following or tracing a stimulus over a period of time [166].

In all cases of subcortical dementia attentional activities, measured with a variety of neuropsychological tests and experimental tasks, are impaired. The only exception is the performance on digit span forward which is unimpaired in PD, HD, MS, SAE and ADC and probably mildly reduced in PSP. Digit span forward is a task which measures immediate memory span, but is often referred to as a test for auditory attentional span [166,278]. Hence auditory attentional span seems to be intact and deficits of attentional activities only become manifest on effort-demanding tasks.

Executive control functions
Impairment of several so-called 'executive' control functions is present in patients with evidence of dementia. Least impairment is reported for patients with ADC, but these patients also had less pronounced general intellectual deterioration.

The presence of these deficits of executive functions and of frontal lobe dysfunction, and also the reduced verbal fluency seems to imply selective involvement of frontal lobe functions in patients with subcortical dementia.

Language
Speech difficulties are present in all seven diseases.

Although slight word-finding difficulties are sometimes present, characteristics of anomic failure as in aphasia syndromes are absent. Verbal fluency is reduced, although little information is available on the performance of patients with NPH. Reduced verbal fluency is also one of the characteristic features of patients with frontal lobe lesions or disease [150]. (Demented) patients with HD, PD, NPH, and SAE sometimes manifest mild comprehension difficulties. For patients with SAE these deficits are probably due to the presence of cortical infarcts.

Based on the literature reviewed, the comprehension difficulties in patients with HD, PD and NPH seem to be manifest only on more complex language-related tasks, e.g. Token test. Performance on these tasks also involves attention and memory [166]. Repetition is unimpaired.

Handwriting can be defective but this constituted a motor impairment, related to the underlying disease. The linguistic quality of writing and reading comprehension are normal. One has to be cautious with these last findings, as the tests normally used are relatively simple and short.

Taken together, a mild language disorder is present, especially with regard to verbal fluency and more complex comprehension tasks, without the characteristics of aphasia.

Arithmetic
Basic arithmetical operations can be performed normally. In mildly demented patients arithmetical deficits become apparent on more complex tasks, e.g. 'serial sevens' and the arithmetic subtest of the WAIS(R). Tasks like 'serial sevens' are also used as measures of attention. Difficulties in immediate memory, concentration or conceptual manipulation and tracking can be the cause of defective performance on arithmetical questions [166,278].

Praxis

Apraxia is defined as a disorder of execution of learned movements which cannot be accounted for by either weakness, incoordination, sensory loss, incomprehension, or inattention to instructions. This disorder is not present, with the exception of so-called 'gait apraxia' in patients with NPH. The term 'gait apraxia' refers to the inability to walk in the absence of primary motor, sensory, or cerebellar deficits and is sometimes seen also in patients with frontal lobe lesions. Gait is a highly complex, sequential motor act. Patients with 'gait apraxia' cannot easily be demonstrated to perform the same motor act in another context. This deficit seems to reflect impairment of motor mechanisms and should better be referred to as 'frontal gait disorder' [152].

Visual information processing

Visual information processing covers a wide range of cognitive abilities including visual recognition, complex visual perception (visual organization, visual interference) and visuospatial ability, e.g. appreciation of the relative position of stimuli, integration of objects within a spatial framework and performing spatial mental operations [33,166,230].

Although oculomotor and other neuro-ophthalmic abnormalities are often present, basic visual gnosis is normal. Minor to mild impairment of complex visual perception is described, although the neuropsychological information with regard to patients with NPH is scarce. Mild to moderately impaired performance on visuospatial tasks is frequently reported. It is conceivable that visuoperceptual deficits are only present on more complex tasks, which also make demands on the capacity for sustained attention and/or ideational-associative capacity [166].

Two recent studies of visuospatial function argued against a generalized visuospatial deficit in PD patients. One study suggested difficulty in utilizing information regarding complex visuospatial tasks [26]. In the other study PD patients were unimpaired in the aspects of spatial function tapped by the task investigated [200]. Based on a review of research on visuospatial function Brown and Marsden (1986) suggested that poor performance of PD patients on many spatial tasks can be more parsimoniously explained by other cognitive factors, e.g. impaired ability to 'shift' a cognitive or motor strategy [36].

Taken together the evidence suggests that visual gnosis is normal, and visuoperceptual and visuospatial impairment is found on tasks requiring attention and/or aspects of executive control function.

Visuoconstruction
Visuoconstructional performance combines perceptual activity with motor response, always has a spatial component, and implies organizing activity [20,166].

(Very) mild impaired performance on drawing tasks is reported. Impairment on tests like block design is more prominent. Defective planning has shown to be the primary cause of the difficulties experienced by demented PD patients on visuoconstructional tasks [73].

It seems plausible that, in combination with the aforementioned difficulties in visual information processing, visual constructional impairment is caused by a combination of cognitive deficits including executive control function.

Motor speed
On clinical examination, motor disturbances are a characteristic feature. Slow performance on tasks measuring motor speed is present in patients with subcortical dementia. Data were lacking for patients with SAE, but based on our own experience the majority of these patients are slow in performing motor tasks [74] (Derix, unpublished observations). Unless accounted for, the motor slowing can be a confounding factor in interpreting results on tasks requiring a motor response.

Subcortical dementia: one syndrome?

Subcortical dementia and frontal lobe disorders
Based on the literature review, impaired performance is found on neuropsychological tasks requiring: attention; effort; cognitive speed; processing, encoding and manipulating information; and executive control functions (*table 2*).

These deficits, in combination with the changes in affect and emotion mentioned earlier, strongly resemble (pre-) frontal syndromes as can be found in patients with cortical frontal lobe lesions.

Signs and symptoms of (pre)frontal lobe syndromes can be grouped under: motor disorders; impairment of attention; changes in high level cognitive ability, including goal-directed motor behaviour; and emotional and affective abnormalities. The appearance of components of each of these depends on the location of the lesion(s) [67,92].

The motor disorder may consist of hypokinesia, e.g. a general diminution of spontaneous motor activity, or of hyperkinesia, e.g. excessive and aimless motor activity [92].

Five categories of attention disorder are distinguished: low general awareness of events in the environment, manifested by reduced drive or interest; impaired awareness of the contralateral side of one's own body; changes in control of eye movement, e.g. disorders of visual search and gaze control; distractibility by irrelevant sensory stimuli and inability to resist this interference; and the inability to concentrate on a given trend of action or thought [67,92,150].

The impairment of cognitive abilities narrows down to disorders of temporal organization or inability to organize new and deliberate sequential behaviour. This results in deficits of memory and planning, and the inability to avoid interference from both internal and external sources. These deficits are related to each other and to attention and drive [67,92,150].

The standard descriptions of affective and emotional changes in frontal lobe patients include an apathetic syndrome and an euphoric syndrome, of which variants can occur in mixed forms [67,92,150].

The similarity between the behavioural and cognitive consequences of frontal lobe lesions and of subcortical dementia syndromes, can be addressed from two (related) viewpoints:
- a model of cognitive function based on the concept of information processing
- neuroanatomical, neurochemical and neuroimaging findings

Information processing model of cognitive function

In contemporary models of cognitive neuropsychology the person is seen as an information processing system. The information processing paradigm is a conceptual approach which assumes that human function can be conceptualized and understood in terms of how both environmental and internal information is processed and used. The focus is largely upon the structures and operations within the system and how they function in selecting, transforming, encoding, storing, retrieving and generating information and behaviour [127]. As such, information processing models are composed of functionally independent knowledge structures connected in a particular organization, in which performance deficits are not related to site of lesion but to a model of cognitive function. This requires that the precise nature of the problem needs to be specified with respect to the underlying organization of the cognitive system. In most models a modular organization of specialized (sub-)systems is regarded as the optimal arrangement for the representation of information. Modules are functional, cognitive processing subsystems. As such the information processing approach can provide a conceptual framework for studying the neurologic basis of psychological processes [80,91]. In figure 1. a simple example of the modularity approach is presented (*figure 1*).

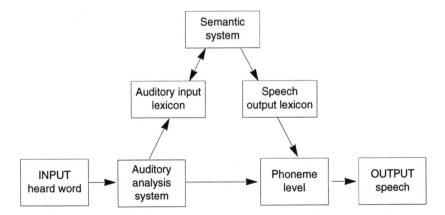

Figure 1. Simple illustration of the modular structure of an information processing model for the recognition, comprehension, and repetition of spoken words. (after Ellis & Young, 1988 [80])

Norman and Shallice's model

A promising recent cognitive neuropsychological model for encompassing a broad range of cognitive functions, is the model of Norman and Shallice [211,253], elaborated by Shallice [252], and based on the concept of modularity (*figure 2*). This model seems particularly suitable for describing not only isolated but also combinations of neuropsychological deficits [252,253].

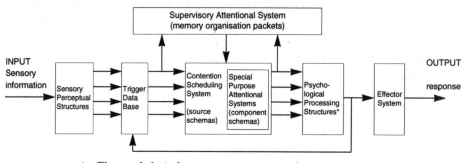

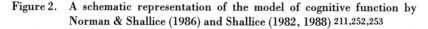

* The psychological processing structures also receive processing input from the sensory perceptual structures

Figure 2. A schematic representation of the model of cognitive function by Norman & Shallice (1986) and Shallice (1982, 1988) [211,252,253]

Terms which will be elucidated in explaining this model are: sensory perceptual structures; psychological processing structures; effector systems; source schemata and component schemata; trigger data base; Memory Organisation Packets in relation to episodic and semantic memory; and three levels of attentional control: special purpose attention control systems, Contention Scheduling System and Supervisory System.

In this model the modules operate in schemata. A *schema* is an independent unit that can control a specific overlearned action or skill, such as answering a phone or doing a long division.

Source schemata, divided in action schemata and thought schemata, are specialized routine programs, each of which will produce a specific output for a certain range of inputs.

Each source schema has a level of activation, dependent on the triggering input it receives from the *trigger data base*. When the level of activation exceeds a given threshold, a schema is selected. The activation level of these schemata can be triggered in various ways, for instance by input from perceptual systems (*sensory perceptual structures*) and/or by the output of other schemata. Once a source schema is selected, it continues to operate to reach its goal or to complete its operations, unless it is actively switched off or blocked.

For the execution of source schemata lower level schemata (*component schemata*), and their related processing (sub)systems (*psychological processing structures*) are called upon.

The execution of schemata can be blocked when necessary further resources or information are lacking, or when the processing (sub)systems are being utilized by some more highly activated schema. An example of the first is the answering the phone. The ringing triggers the routine action "answering the phone"; when, after taking the receiver off the hook and saying your name, no reaction follows, you will discontinue this action.

The selection of a source schema frequently requires that the variables connected with the goal of the schema are set, e.g. catching a ball requires the 'catching' action to be elicited for the expected position of the ball. Selection of lower level or component-schemata leads to a particular arrangement of trans-mission routes between processing subsystems being switched in. Also, the selection of (component) schemata implies that the various sub-routines and processing (sub)systems are activated. This leads to a particular arrangement of transmission routes between processing subsystems switched in. Finally, the selected component-schemata may need to adjust the operations of the processing subsystems which they utilize. This is done by *special purpose attentional control systems*, often controlled by the source schemata themselves. This kind of control is required in situations in which there are groups of processing subsystems that operate in a fairly tight interconnected and fixed fashion, but in which control problems can arise because individual subsystems have more inputs than they can process at a time or because the different subsystems operate at different speeds. An example of a special purpose attentional control system is the visuospatial attention system.

Finally, if a kind of external response is required, this will be effected by employing the relevant *effector system(s)*.

More than one independent schema is capable of controlling the processing (sub)systems it requires at the same time. Schemata are independently triggered and are in mutually inhibitory competition for selection. Which schemata will be inhibited and by how much, depends on the particular processing (sub)systems they require. The control of routine selection of the routine actions and/or thought operations (source schemata) is decentralized, and this process is termed Contention Scheduling.

The Contention Scheduling System is the second attentional control system. It exerts a lateral activation and inhibitory control on the operations of the source schemata involved. This system ensures, in a routine way, the efficient use of the limited effector and cognitive resources. The Contention Scheduling System resolves competition for selection among activated schemata by preventing competitive use of common or related processing (sub)systems. It also takes care of shared use of common structures or operations when possible.

The existence of special purpose attentional control systems and the Contention scheduling system alone are insufficient for explaining all levels of selections of thought and action operations.

A three-level control structure for the regulation of sub-system behaviour is proposed, in which a general-purpose supervisory system, the Supervisory Attentional System or *Supervisory System* is the third important component [211,252,253]. By including this system, the model of Norman and Shallice [211] can be considered the information processing realisation of the theory of Luria [252].

The Supervisory System corresponds to Luria's functional system for the programming, regulation and verification of activity, related to the organization of conscious activity, and residing in the frontal lobes [174]. This high level supervisory attentional control system has limited capacity and is internally modular – different functions can be affected by damage to different subcomponents.

The Supervisory System is involved in the genesis of willed actions, and is necessary for satisfactory performance in non-routine tasks, in which it has to produce a response to novelty that is planned rather than one that is routine or impulsive. This system has access to a representant of the environment and of the person's intentions and cognitive capacities. These are the higher level schemata or programs called *Memory Organisation Packets* (MOPs).

MOPs are overall organisations of (complex) activities in which the actions to be undertaken may be realized in a variety of ways. MOPs are selected under the control of the Supervisory System, and make use of the information stored in episodic memory [252].

Episodic memory contains the autobiographical record of the person's previous environments, activities, plans, and intentions. Lower level 'episodic' information is laid down in the same subsystems that carry out on-line semantic processing, which are controlled by the Contention Scheduling System.

Semantic memory can be viewed as a person's knowledge about the meaning of words and other verbal symbols and facts, but also non-verbal knowledge, such as the significance of objects or symbols. Information in the semantic memory system is accessible by the operations of some routine schema (Contention Scheduling) operating directly on the processing systems. Access to information in the episodic memory system requires the formulation of a description by the Supervisory System and verification of any record retrieved [252].

By selecting MOPs for particular situations, the Supervisory System operates by adjusting the Contention Scheduling System. It exerts this influence by way of activating or inhibiting particular source schemata that directly control (routine) actions and thought operations. In this way, it can override automatic and routine action or processing (Contention Scheduling System) and focus attention elsewhere, depending on the nature of incoming information and the existing MOPs with regard to the situation at hand. The Supervisory System gets information from the trigger data base and from the (results of) the operations of the Contention Scheduling System [252].

Damage of the Supervisory System results in reliance on the Content Scheduling system alone, and gives rise to symptoms associated with frontal lobe disorders. These consist of: distractibility and a tendency to be side-tracked by irrelevant associations, c.q. impairment on tasks requiring concentration; behaviour rigidity – a tendency to perseverate – because the schema controlling routine action or thought operation remains fixed, without there being a malfunction of the subsystems being used to carry out the schema; and defective planning, regulation and verification of behaviour, with consequences for problem solving behaviour and memory function [252].

Impairment of the psychological processing structures, incorporated in the operations of the Contention Scheduling System, will give rise to specific cognitive disorders of routine actions or thought operations, e.g. apraxia, acalculia, agnosia or aphasia.

Based on literature with regard to the behavioural effects of psychomotor drugs, Norman and Shallice (1986) suggested that changes in dopaminergic input to the basal ganglia can influence the activation level of schemata. Increased dopaminergic input can result in a breakdown of the lateral inhibitory control of the Contention Scheduling System. Of importance is the possible effect of decreased dopaminergic input, which can result in a complementary condition of lowered activation levels of source and component-schemata [211]. This would imply that it takes longer before the activation level of a schema exceeds a given threshold.

Damage of the special purpose attentional systems and their lower level, component schemata can result in specific cognitive sensory and motor deficits, as for instance attentional dyslexia [252].

In table 3 the functions and associated cerebral systems of the Contention Scheduling System and the Supervisory System are summarized (*table 3*).

Table 3.   Contention Scheduling System and Supervisory Attentional System: functions and cerebral structures involved.

	CONTENTION SCHEDULING SYSTEM	SUPERVISORY ATTENTIONAL SYSTEM
CONTROLLING FUNCTION	routine actions, and thought operations by using source-schemas*	non-routine tasks , and willed actions by using MOP's
CEREBRAL STRUCTURES	association cortices, memory structures, basal ganglia	(pre-)frontal cortex

*   The necessary component-schemas are under the control of special purpose attentional systems

Damage to the sensory perceptual structures and the effector systems will result in focal (neurological) deficits like for instance visual field defects, paresis or peripheral motor deficits.

Subcortical dementia and the model of Norman and Shallice
The cognitive deficits in the described subcortical dementia syndromes, seem at first sight similar to frontal lobe disorders, but there are some differences in degree.

The main emphasis in the early stage of subcortical dementia is on a decline of efficient processing and encoding new information and of utilizing (manipulating) old and new information, which manifests itself in more complex situations requiring attention and effort. Attentional deficits like distractibility are only found on effort-demanding tasks. General rigidity of behaviour and prominent general deficits of planning, regulation and verification of behaviour have not been found; deficits, if present, are task specific.

The neuropsychological deficits in subcortical dementia can be interpreted as the result of impaired access to the Supervisory system itself, of impaired access of the Supervisory system to the operations of the Contention Scheduling System, and a change in the activation level of the schemata and component-schemata, most probably a lowered level of activation. In the first place, if the total resource of the limited capacity of the Supervisory System can not be adequately used, performance will decline when the task demands increase beyond the available resources of this system. This decline will manifest itself primarily in an increase of time needed for performance, the inability to finish the task, and sometimes in faulty performance, e.g. relying on a routine program not adequate for the specific situation. Secondly, for efficient operation of the Supervisory system it requires selecting of relevant MOPs. If this information cannot be adequately accessed, deficits in the necessary adjustment of the operation of the Contention Scheduling will occur. This will result in (partially) incorrect performance, e.g. selection of an inadequate program, and also an increase in response-time. Accepting that the psychological processing structures, incorporated in the operations of the Contention Scheduling are intact, no deficits, except a possible slowed performance, should be found on tasks requiring only routine selection between routine actions or routine thought operations.

The results of the literature review on neuropsychological deficits in subcortical dementia are consistent with this. Recent findings in a study of patients with PD give extra support. The performance of PD patients was impaired on attentional tasks which required internal attentional control but not on tasks which provided external control [34]. PD patients were able to execute a given plan of action, were able to generate low level strategies, required for efficient searching, and had normal spatial working memory capacity compared with matched controls. But on a problem solving task deficient planning was found, measured as an increase in time needed to think before executing the first move [200].

Depending on the severity of both kinds of impaired access to and from the Supervisory system, the performance impairment would be more compromised. In less severe cases of subcortical dementia, the performance deficits would be less specific and manifest itself on a variety of only complex and/or effort demanding tasks. In more severe cases, with more extensive cerebral (subcortical) involvement, it seems logical to expect cognitive deficits more and more similar to symptoms of frontal lobe disorders.

Neuroanatomical and related findings in subcortical dementia

All neurological disorders described in this chapter are characterized by damage to subcortical regions, although the site of and the extent to which subcortical grey structures and white matter are involved differ, as described in the beginning of this chapter.

NPH, SAE, MS and ADC can be considered 'white-matter dementias', in which one or more white matter pathways can be damaged [82]. In NPH the ventricles are enlarged and adjacent periventricular gliosis is present. Some evidence of deep grey matter ischemia is present in SAE, but the more prominent feature is diffuse white matter loss along the lateral ventricles. Cerebral demyelination of the central and periventricular matter is the striking finding in AIDS patients with ADC. The burden of white matter involvement in MS appears to fall heavily on frontal lobe white matter [82].

Different subcortical structures are affected in PSP, HD and PD. PSP is characterized by severe neuronal loss in the striatum and substantia nigra, associated with degeneration of other structures in the basal ganglia, thalamus, brainstem and cerebellum. Lesions in HD are predominantly found in the striatum, particularly in the caudate nucleus and pallidum; some damage of thalamus is possible; only in later stages the damage extends to the limbic system, cerebral cortex and cerebellum. In PD severe neuronal loss occurs in the substantia nigra, although in later stages a limited loss of intracortical neurons can be found [2].

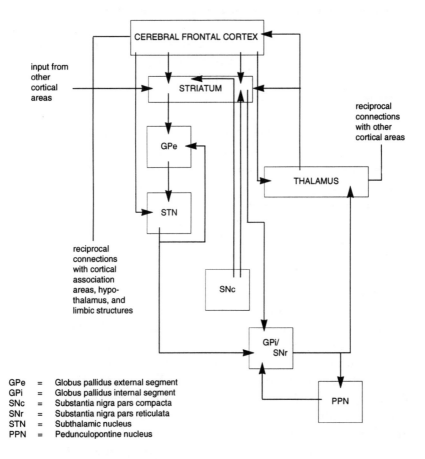

GPe = Globus pallidus external segment
GPi = Globus pallidus internal segment
SNc = Substantia nigra pars compacta
SNr = Substantia nigra pars reticulata
STN = Subthalamic nucleus
PPN = Pedunculopontine nucleus

Figure 3.   Basal ganglia-thalamocortical circuitry [7,8,92,194] (adapted from Alexander & Crutcher 1990 [7])

In one way or another, damage of these subcortical structures and/or cerebral pathways will exert its influence on the function of the prefrontal cortex.

The prefrontal cortex has reciprocal connections with: cortical sensory association areas; the thalamus; the hypo-thalamus; limbic structures (hippocampus and amygdala); and the midbrain (reticular activating system and substantia nigra). The thalamus also conveys influences from lower levels of the brainstem and limbic structures (e.g. amygdala) to the (pre)frontal cortex, and has reciprocal connections with other cortical areas. The efferent projections of the prefrontal cortex to the basal ganglia, notably the caudate nucleus of the striatum, are unidirectional 92,194. The striatum receives also projections from other cortical areas, the thalamus, substantia nigra, globus pallidus; it projects only to the globus pallidus and substantia nigra 194,213.

There are several (in)direct connections between subcortical structures and the prefrontal cortex, which are part of the basal ganglia-thalamocortical circuits (*figure 3*). Five basal ganglia-thalamocortical circuits have been suggested, each receiving multiple partially overlapping cortico-striate inputs 7. These are: the motor circuit which focuses on the precentral motor fields; the oculomotor circuit with its focus on the frontal and supplementary eye fields; two prefrontal circuits respectively projecting to the dorsolateral prefrontal cortex and the lateral orbitofrontal cortex; a limbic circuit which focuses on the anterior cingulate cortex and medial orbitofrontal cortex 7,8 (*figure 4*). The inputs of these circuits are progressively integrated in passing through the pallidum and substantia nigra and project to the thalamus and from there to single cortical areas, especially the prefrontal cortex 2,92,98,194. Specific disturbances within the basal ganglia-thalamocortical 'motor' circuit lead to specific movement disorders 6,71,213. In these circuits, the thalamus serves as relay station for all sensory pathways to the cerebral cortex 194. The anterior nucleus of the thalamus is connected with the cortex of the gyrus singuli. The medial nuclei are connected with the orbitofrontal and prefrontal cortex; the ventral anterior and ventral lateral nuclei are connected with the more rostral parts of

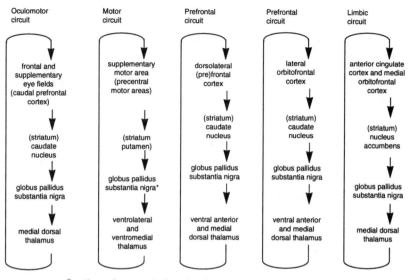

Oculomotor circuit	Motor circuit	Prefrontal circuit	Prefrontal circuit	Limbic circuit
frontal and supplementary eye fields (caudal prefrontal cortex)	supplementary motor area (precentral motor areas)	dorsolateral (pre)frontal cortex	lateral orbitofrontal cortex	anterior cingulate cortex and medial orbitofrontal cortex
(striatum) caudate nucleus	(striatum putamen)	(striatum) caudate nucleus	(striatum) caudate nucleus	(striatum) nucleus accumbens
globus pallidus substantia nigra	globus pallidus substantia nigra*	globus pallidus substantia nigra	globus pallidus substantia nigra	globus pallidus substantia nigra
medial dorsal thalamus	ventrolateral and ventromedial thalamus	ventral anterior and medial dorsal thalamus	ventral anterior and medial dorsal thalamus	medial dorsal thalamus

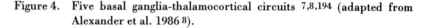

* there also is an 'indirect' pathway via putamen → external segment of globus pallidus → subthalamic nucleus → internal part of the globus pallidus/substantia nigra pars reticulata → thalamus.

Figure 4. Five basal ganglia-thalamocortical circuits [7,8,194] (adapted from Alexander et al. 1986 [8]).

the frontal lobe, respectively area 6 and 4. The ventral posterior thalamic nucleus is joined to the postcentral gyrus. The thalamus is also connected within the limbic system and the brain stem reticular activating system [194,195,210,213]. The reciprocal connections with limbic and reticular activating system imply an important role for the prefrontal lobe in the modulation of arousal, motivation and affect.

Within the prefrontal lobe the different cortical areas differ in their afferent and efferent connections, which explains the different behavioural manifestations of (localised) frontal lobe lesions with regard to: attention; motility (hypokinesis or hyperkinesis); temporal integration of behaviour (memory, planning and interference control); and mood and affect [92,150,194,195].

An implication of the concept of basal ganglia-thalamocortical circuits is that disruption at different points within a circuit might be expected to have similar behavioural consequences [32].

It seems reasonable to presume that, even when the prefrontal cortex is intact, damage of subcortical structures and/or prefrontal afferent and efferent, direct and indirect, connections will influence cognitive functioning, emotional behaviour and motor function.

The behavioural and cognitive deficits in subcortical dementia seem thus to be caused by a disturbance of prefrontal cortex function, and by a disturbance of the activation levels of different cortical functions, depending on the cerebral structures and pathways involved.

Evidence for this also comes from neurochemical findings. Damage of subcortical structures results in neurotransmitter abnormalities in subcortical structures and in related cerebral pathways, which give rise to motor disorders and behavioural abnormalities [6,71,236]. It has been shown that cholinergic pathways in the brain determine the efficiency of attentional processing and are also involved in memory function together with a noradrenergic system [276]. Noradrenergic pathways are also involved in discriminative selectivity of responding. Depletion of dopamine leads to a deficit in the ability to initiate, but not to a deficit in the ability to execute an action [236].

Evidence for cortical (often frontal) dysfunction in patients with subcortical structural lesions also comes from neuroimaging studies. A study of brain function with PET-scanning in patients with PSP showed a global decrease in blood flow and oxygen metabolism, which was most marked in the frontal regions; also striatal dopamine formation and storage was significantly decreased, and this decrease paralleled the degree of frontal cerebral blood flow hypofunction [162]. In a recent study of patients with moderate to severe HD, cerebral metabolic rate of glucose consumption measured with PET-scan, was reduced in the striatum and in the cortex, most markedly in the frontal cortex [159]. PET-scan studies in patients with PD have shown decreased values of cerebral blood flow and metabolism in the basal ganglia and the frontal cortical regions [216,285]. Studies with regard to cerebral metabolic function in patients with NPH were scarce. In a PET-scan study of three patients a general cortical hypometabolism was found. The authors remarked that the procedure used did not permit to detect changes in periventricular white matter [129]. No PET-scan studies of patients with SAE were found. Metabolic dysfunction has been shown in patients with ADC using PET-scan. In early ADC subcortical hypermetabolism was present. As the disease progressed subcortical grey matter (thalamus and basal ganglia) and cortical hypometabolism, including the frontal cortex, was found [240], although the pathology in these patients predominantly involves subcortical structures, with the cortex being only mildly affected [204]. One study measuring brain function in patients with MS was found. It was shown that cerebral oxygen utilization and blood flow were significantly reduced in MS patients, both in the cerebral white matter and peripheral cortical grey matter [31].

Discussion and conclusions

There is considerable overlap between the cognitive and emotional disturbances in the described subcortical dementia syndromes. In the early stages of subcortical dementia cognitive impairment consists of a decline of efficient processing and encoding new information and of manipulating old and new information, which manifests itself in more complex situations requiring attention and effort. There are similarities with the various behavioural consequences of frontal lobe lesions, including changes in mood and affect.

The model of cognitive function of Shallice [252] provides an explanation of these disorders. This information processing model, is based on a three level control structure for the regulation of subsystem operation: local, special purpose, attentional control systems; the contention scheduling system; and the supervisory system. Basic units underlying action or thought are discrete programs (schemata), that can control a specific overlearned action or skill, divided into:

1. routine programs (source schemata), composed of component schemata, which are under the control of respectively the contention scheduling system and special purpose attentional control systems;
2. higher level programs (MOPs) which are under the domain of the Supervisory System. The system of Contention Scheduling

provides for the decentralized routine selection between routine actions or thought operations. The Supervisory System is required in the genesis of willed actions and in non-routine selection, meaning situations where the routine selection of actions is insufficient to cope with the demands of the situation. Executing MOPs requires access to the system of Contention Scheduling.

Adequate execution of routine actions or thought operations involves the association cortices and cerebral memory structures. Basal ganglia function influences the activation level of these schemata. Damage of the psychological processing structures necessary for executing the routine actions or thought operations results in cognitive disorders like aphasia and apraxia. Supervisory control is executed by the frontal lobes, and damage of the Supervisory system results in frontal lobe disorders.

The main disturbance in the early stages of subcortical dementia seems impaired access to the Supervisory System or impaired access of this system to the Contention Scheduling system, compromised by a decreased activation level of the schemata. This results in less efficient, slowed performance with the possible occurrence of errors on more complex, more effort demanding cognitive tasks. Routine actions and thought operations, although sometimes with a delay, can be performed as they depend on content scheduling alone. In later stages of subcortical dementia, depending on the extent of the damage of access to and from the Supervisory System, clear frontal deficits can appear.

The neuroanatomical, neurochemical and neuroimaging findings in diseases with subcortical pathology are consistent with the notion of deafferentiation of the cortex, generally the prefrontal lobes. In the so-called 'white matter dementias', in which not only subcortico-cortical but also cortico-cortical connections may be damaged, structural lesions of cortical circuits may also be present.

Albert et al (1974) initially proposed as a tentative hypothesis that the common mechanisms underlying subcortical dementia are those of impaired timing and activation of normal intellectual processes [5]. The impaired function of the reticular activating systems or a disconnection of the reticular activating systems from thalamic and subthalamic nuclei was thought to result in a slowing down of normal intellectual processes. They speculated that the cortical systems responsible for perceiving, storing and manipulating knowledge were relatively intact. They may however be abnormally activated; and, once activated, they may take an excessive amount of time to carry out intellectual processing. In their publication they elaborated on the similarity of clinical characteristics between frontal lobe syndromes and subcortical dementia, and the anatomical connections between the frontal lobe and subcortical structures. A decade later they proposed to replace the term subcortical dementia by the term fronto-subcortical or frontal system dementia [89].

Based on the present evidence this explanation is still viable, but the different cognitive deficits described, seem better explained in terms of the information processing model of cognitive function developed by Shallice [252]. Damage of subcortical structures and/or their connections with the reticular activating system can be seen as leading to a disturbance of activation levels of (component and source) schemata, most often resulting in slowed down performance. Damage of basal ganglia-thalamo-cortical circuits, and especially those in which the (pre)frontal cortex is involved, results in impaired function of the Supervisory System. This will manifest itself on neuropsychological tasks requiring effort and the performance of executive control functions. The extent of impaired performance depends on the subcortical structures and cerebral circuits involved.

Thus it can be proposed that subcortical dementia syndromes are characterized by a pattern of cognitive deficits which can be explained by common underlying mechanisms. In order for this proposition to be relevant, it must be shown that this pattern is distinct from that of cortical dementia syndromes, and that the latter are explained by other mechanisms.

The possible role of the model of cognitive function of Shallice [252] in cortical dementia can be inferred from Alzheimer's disease (AD) which is the main cause of cortical dementia [59]. The clinical signs and symptoms concern the so-called 'instrumental' functions and consist of progressive memory disturbances, aphasia, acalculia, apraxia, impaired visuospatial skills and personality changes [54]. Instrumental functions are dependent on the integrity of the cerebral (association) cortex and hippocampus [54]. The cognitive disorder in AD is indicative of impairment of the psychological systems involved in the operations of the Contention Scheduling system, resulting in so-called higher cortical deficits. These deficits will also influence the (outcome of) operations of the Supervisory System. Neuropathological findings and results of neuroimaging studies are in accordance with this explanation. AD is neuropathologically characterized by widespread cortical senile plaques and neurofibrillary tangles, with the exception of the primary sensory and motor cortices. Neurofibrillary tangles are found particularly in the association areas. Probably there is also some cortical neuronal loss. Widespread damage to the ascending cholinergic system is present; although this also includes the primary motor and sensory cortices, it does not result in motor weakness or sensory disturbances [59,239]. This suggests that motor and sensory deficits do not appear unless additional cortical pathology is present, as is the case in the association areas [59,239]. Neuroimaging studies of cerebral blood flow and metabolism have particularly shown reduced levels in the posterior temporal and parietal cortical regions, significantly more severe than in the frontal cortex [22,64,90,119,129].

Various studies have repeatedly reported neuropsychological differences and/or different neuropsychological profiles of cognitive impairment in patients with AD when compared with demented patients with PD [48,61,86,87,88,113,114,122,126,219,222,261], when compared with demented patients with HD [29,41,42,113,143,202,219,247], and also when compared with patients with PSP [196,219,222]. An often-quoted study [186] failed to reveal a (cognitive) distinction between patients with 'cortical' and 'subcortical' dementia based on the mini-mental state examination (MMSE). In this study the MMSE was used both to define the level of severity, and for intergroup qualitative comparisons. However, it has been repeatedly reported that this instrument is unsuitable for providing detailed information with regard to individual cognitive functions and is therefore inadequate for making qualitative comparisons [126].

As already touched on in the Introduction, the classification in cortical and subcortical dementia syndromes is controversial, and has proponents and opponents. Part of this problem seems to stem from the nomenclature, which seems to imply a strict anatomical distinction, although the authors who introduced [5] and consequently propagated [59,63] these concepts, insist that it is a clinical classification. Subcortical dementia is a clinical syndrome resulting from dysfunction of mainly subcortical structures and/or white matter tracts. The clinical syndrome of cortical dementia is found in diseases with mainly, and prominent cortical changes.

A second, related, problem concerns the fact that diseases classified as subcortical dementias may have cortical pathology and vice versa, such as the frequently mentioned lesions in the nucleus basalis of Meynert in patients with AD [280] and Alzheimer-type lesions in the cortex of patients with PD [25]. The significance of the changes in the nucleus basalis in AD is not yet clear [239]. Recently a subgroup of demented subjects has been described with the neuropathological characteristics of AD but with a relatively spared basal nucleus [282]. Neuronal loss in the nucleus basalis of Meynert has been found in some patients with PD and dementia but the relationship between this finding and the dementia has not been resolved yet [55,60]. It has been proposed that a clinical distinction, related to the neuropathological findings, can be made between dementia in PD mainly caused by a dopaminergic deficiency (and sometimes a superimposed cholinergic deficiency), and dementia caused by the concurrence of PD and AD [60]. In Parkinson's dementia the main changes are neuronal degeneration in basal ganglia, amygdala and thalamus; in PD plus AD (PD+AD) the same pattern of subcortical degeneration is present, but combined with degeneration in the cerebral cortex, and hippocampus [69].

A third problem with regard to the distinction between cortical and subcortical dementia is the frequent overlap in clinical findings between patients with cortical and subcortical dementia syndromes. Extrapyramidal signs, particulary rigidity and tremor, have been reported in a proportion of patients with dementia of the Alzheimer type [185]. However, in a study with positron emission tomography, comparing AD patients with and without extrapyramidal signs and a control group, no differences were observed with regard to the fluoro-18-dopa uptake into caudate nucleus and putamen, in contrast to the marked reduction in the putamen observed in PD. The authors suggest that extranigral factors may be involved in the pathogenesis of rigidity in AD [270].

A fourth problem regarding the strict adherence to two superordinate categories of dementia concerns differences between dementia syndromes belonging to one of these categories. As already mentioned, within the overall picture of a subcortical dementia, subtle differences have been found between the performances of patients with dementia due to different, mainly subcortical, diseases. Within the overall pattern of (cortical) deficits in patients with Alzheimer's disease, distinct subgroups of patients with AD have been reported [209], characterized qualitatively by different profiles of cognitive impairment and corresponding patterns of cerebral hypometabolism, studied with positron emission tomography [179]. Non-Alzheimer type dementia associated with cerebral atrophy with prominent amnesia and apraxia or with a combination of amnesia, a perceptuo-spatial disorder and aphasia has also been described [209]. A syndrome of progressive aphasia has been reported in patients in whom general cognitive ('cortical') impairment became only manifest several years after the onset of the language impairment [108,153,223]. Postmortem neuropathological examinations are rare and many causes are found: Pick's disease, Alzheimer's disease, or aspecific cortical gliosis [57,108,193,199]. A progressive dementia has been described which was heralded by alexia or visual agnosia, with visuoconstructional impairment, and in which all five patients eventually developed alexia, agraphia, visual agnosia, acalculia, finger agnosia, right-left disorientation and transcortical sensory aphasia; memory, insight and judgment were relative preserved until late in the course. Computed tomography and magnetic resonance imaging revealed a degenerative process prominent in the posterior cerebral hemispheres in three of the patients. The EEG of all patients showed patterns resembling those reported in the mild stages of both Pick's disease and Alzheimer's disease [17]. A similar patient is described by Croisile et al (1991) [58]. Within the model of Shallice [252] these patterns of cognitive deficits point to impairment of the Contention Scheduling System. Pick's disease, a dementia with predominantly frontotemporal cortical degeneration, is also classified as a 'cortical' dementia, but there are differences in clinical phenomenology when compared with AD; changes of mood, and abnormal behaviour are among the earliest signs in Pick's disease and amnesia, visuospatial disorientation and acalculia occur only in later stages; the language impairment in early Pick's disease is relatively minor, with semantic anomia and circumlocution, and is reminiscent of frontal lobe impairment [59,155,267]. In recent years a frontal lobe dementia of non-Alzheimer type has been described, in which unspecific frontal or frontotemporal grey matter changes are found, but not the circumscribed atrophy seen in Pick's disease, although the clinical differentiation from Pick's disease can be difficult [154]. The typical clinical picture of dementia of (non-Alzheimer) frontal lobe type consists of slowly progressive dementia with stereotyped speech, relatively spared receptive speech function, variable memory impairment, social breakdown and personality changes with lack of concern and insight, and absence of visuo-spatial disorder [109,207].

Based on the model of Shallice [252], it seems logical to expect a different profile of cognitive dysfunction in cortical dementia with predominantly frontal lobe involvement. In these cases it appears that the main damage concerns the Supervisory System, resulting in frontal lobe symptoms. Alzheimer type dementia with striking frontal-lobe type symptomatology has also been described, but the symptom profile in these patients can be distinguished from the cognitive deterioration in Pick's disease and frontal-lobe-type dementia, because of the prominent presence of cognitive deficits related to temporo-parieto-occipital cortical dysfunction [38]. Within the model of Shallice [252] this suggests damage of the Contention Scheduling System as well as the Supervisory System. As already mentioned, symptoms suggesting frontal lobe dysfunction have been found in patients with mainly subcortical degenerative diseases leading to dementia, especially progressive supranuclear palsy [219,222].

Until now no studies were found comparing neuropsychological performance of patients with ((non-)Alzheimer) dementia of the frontal lobe type and patients with 'subcortical' dementia. As already mentioned there seem to be differences in degree between the cognitive deficits in frontal lobe disorders and those in subcortical dementia. Also, differences have been found with single photon emission tomography (SPECT), between patients with a clinical diagnosis of AD, non-Alzheimer frontal-lobe dementia and PSP [208]. Posterior hemisphere abnormalities were common in the Alzheimer group, rare in non-Alzheimer frontal-lobe dementia, and absent in patients with PSP. Selective abnormalities in the anterior hemispheres were characteristic of these last two groups, and occurred only rarely in the Alzheimer group. Lastly, PSP could be differentiated from the non-Alzheimer frontal-lobe dementia, in that the cortical rim in PSP was intact [208]. In future research attempts should be made to validate this distinction by means of neuropsychological investigation, distinguishing between cognitive deficits caused by damage of the Supervisory System itself, and deficits caused by damaged access to this system. Summarizing: a distinction between cortical versus subcortical dementia syndromes does not mean that all cortical dementia syndromes or all subcortical dementia syndromes are exactly the same with regard to the clinical manifestation and related findings [179,209,219].

The *clinical differentiation* between these two categories seems to hold, despite the frequency of neuropathologically and clinically overlapping findings. The different cognitive profiles found in dementia syndromes due to predominantly cortical degenerative diseases still conform to the clinical picture of cortical dementia. Subtle differences have been found within the overall category of subcortical dementia, but these dementia syndromes can still be differentiated from cortical dementia syndromes.

Taken together, based on the neuropsychological, neuro-anatomical, neurochemical and neuroimaging findings, conceptually supported by the model developed by Norman and Shallice [211,253], extended by Shallice in 1988 [252], the concept of subcortical dementia appears to be valid. It describes a clinical syndrome in patients with predominantly subcortical pathology such as in PSP, HD, PD, NPH, SAE, ADC and MS. Subcortical dementia can be differentiated from cortical dementia, not only based on clinical symptoms and results of ancillary investigations but also in terms of cognitive dysfunction based on the model of Shallice.

The results of neuropathological, neurochemical and neuroimaging studies have shown differences with regard to the subcortical structures and cerebral pathways involved in various subcortical dementias.

It is plausible that in the future, with the use of more sophisticated, experimental cognitive tasks in combination with refined neuroimaging techniques it will be possible to find subtle differences between the cognitive symptoms of the various dementing diseases predominantly affecting subcortical structures.

References

1. Adams RD, Fisher CM, Hakim S, et al.
   Symptomatic occult hydrocephalus with 'normal' cerebrospinal fluid pressure. A treatable syndrome.
   N Eng J Med 1965; 273: 117-26

2. Agid Y, Ruberg M, Dubois B, Pillon B.
   Anatomoclinical and biochemical concepts of subcortical dementia.
   In: Stahl SM, Iversen SD, Goodman EC (Eds). Cognitive neurochemistry.
   Oxford, Oxford University Press, 1987: 248-71

3. Albert MS, Butters N, Brandt J.
   Patterns of remote memory in amnesic and demented patients.
   Arch Neurol 1981; 38; 495-500

4. Albert MS, Butters N, Brandt J.
   Development of remote memory loss in patients with Huntington's disease.
   J Clin Neuropsychol 1981; 3(1): 1-12

5. Albert ML, Feldman RG, Willis AL.
   The 'subcortical dementia' of progressive supranuclear palsy.
   J Neurol Neurosurg Psychiatr 1974; 34: 121-30

6. Albin RL, Young AB, Penney JB.
   The functional anatomy of basal ganglia disorders.
   Trends Neurosci 1989; 12(10): 366-75

7. Alexander GE, Crutcher MD.
   Functional architecture of basal ganglia circuits: neural substrates of parallel processing.
   Trends Neurosci 1990; 13(7): 266-71

8. Alexander GE, DeLong MR, Strick PL.
   Parallel organization of functionally segregated circuits linking basal ganglia and cortex.
   Ann Rev Neurosci 1986; 9: 357-81

9. Anzola GP, Bevilacqua l, Cappa SF, et al.
   Neuropsychological assessment in patients with relapsing-remitting multiple sclerosis and mild functional impairment: correlation with magnetic resonance imaging.
   J Neurol Neurosurg Psychiatr 1990; 53: 142-45

10. Babikan V, Ropper AH.
    Binswanger's disease: a review.
    Stroke 1987; 18: 2-12

11. Bamford KA, Caine ED, Kido DK, et al.
    Clinical-pathologic correlation in Huntington's disease.
    Neurol 1989; 39: 796-801

12. Bayles KA, Tomoeda CK.
    Confrontation naming impairment in dementia.
    Brain and Language, 1983; 19: 98-104

13. Beatty WW, Goodkin DE.
    Screening for cognitive impairment in multiple sclerosis. An evaluation of the Mini-Mental State Examination.
    Arch Neurol 1990; 47: 297-301

14. Beatty WW, Goodkin DE, Hertsgaard D, Monson N
    Clinical and demographic predictors of cognitive performance in multiple sclerosis.
    Arch Neurol 1990; 47: 305-08

15. Beatty WW, Goodkin DE, Monson N, Beatty PA.
    Cognitive disturbances in patients with relapsing remitting multiple sclerosis.
    Arch Neurol 1989; 46: 1113-19

16. Beatty WW, Goodkin DE, Monson N, et al.
    Anterograde and retrograde amnesia in patients with chronic
    progressive multiple sclerosis.
    Arch Neurol 1988; 45: 611-19

17. Benson DF, Davis J, Snyder BD.
    Posterior cortical atrophy.
    Arch Neurol 1988; 45: 789-793

18. Benson DF.
    Hydrocephalic dementia.
    In: Frederiks JAM (Ed). Handbook of clinical neurology, vol 2.
    Amsterdam, Elsevier Science Publ BV, 1985: 323-33

19. Benson DF.
    Parkinsonian dementia: cortical or subcortical?
    In: Hassler RG, Christ JF (Eds). Advances in neurology, vol 40.
    Parkinson-specific motor and mental disorders.
    New York, Raven Press, 1984: 235-40

20. Benton A.
    Visuoperceptual, visuospatial, and visuoconstructive disorders.
    In: Heilman KM, Valenstein E (Eds). Clinical neuropsychology, 2nd ed.
    New York, Oxford University Press, 1985: 151-85

21. Berent S, Giordani B, Lehtinen S, et al.
    Positron emission tomographic scan investigations of Huntington's
    disease: cerebral metabolic correlates of cognitive function.
    Ann Neurol 1988; 23: 541-46

22. Berg G, Grady CL, Sundaram M, et al.
    Positron emission tomography in dementia of the Alzheimer type.
    Arch Intern Med 1986; 146: 2045-49

23. Blonder LX, Gur RE, Tur RC, et al.
    Neuropsychological functioning in hemiparkinsonism.
    Brain and Cognition 1989; 9: 244-57

24. Bloxham CA, Dick DJ, Moore M.
    Reaction times and attention in Parkinson's disease.
    J Neurol Neurosurg Psychiatr 1987; 50: 1178-83

25. Boller F, Mizutani T, Roessmann U, et al.
Parkinson's disease, dementia, and Alzheimer's disease:
clinicopathological correlations.
Ann Neurol 1980; 7: 329-335

26. Bradley VA, Welch JL, Dick DJ.
Visuospatial working memory in Parkinson's disease.
J Neurol Neurosurg Psychiatr 1989; 52: 1128-35

27. Brainin M, Goldenberg G, Ahlers C, et al.
Structural brain correlates of anterograde memory deficits in multiple
sclerosis.
J Neurol 1988; 235: 362-65

28. Brandt J, Butters N.
The neuropsychology of Huntington's disease.
Trends Neurosci 1986; 9 (3): 118-20

29. Brandt J, Folstein SE, Folstein MF.
Differential cognitive impairment in Alzheimer's disease and
Huntington's disease.
Ann Neurol 1988; 23: 555-61

30. Brooks DJ, Beaney RP, Powell M, et al.
Studies on cerebral oxygen metabolism, blood flow, and blood volume,
in patients with hydrocephalus before and after surgical
decompression, using positron emission tomography.
Brain 1986; 109: 613-28

31. Brooks DJ, Leenders KL, Head G, et al.
Studies on regional cerebral oxygen utilisation and cognitive function in
multiple sclerosis.
J Neurol Neurosurg Psychiatr 1984; 47: 1182-91

32. Brown RG, Marsden CD.
Cognitive function in Parkinson's disease: from description to theory.
Trends Neurosci 1990; 13(1): 21-29

33. Brown RG, Marsden CD.
'Subcortical dementia': the neuropsychological evidence.
Neurosci 1988; 25: 363-87

34. Brown RG, Marsden CD.
    Internal versus external cues and the control of attention in
    Parkinson's disease.
    Brain 1988; 111: 323-43

35. Brown RG, Marsden CD.
    Neuropsychology and cognitive function in Parkinson's disease: an overview.
    In: Marsden CD, Fahn S (Eds). Movement disorders, vol 2.
    London, Butterworth, 1987: 99-123

36. Brown RG, Marsden CD.
    Visuospatial function in Parkinson's disease.
    Brain 1986; 109: 987-1002

37. Brown RG, Marsden CD.
    How common is dementia in Parkinson's disease.
    Lancet 1984; ii: 1262-65

38. Brun A, Gustafson L.
    Psychopathology and frontal lobe involvement in organic dementia.
    In: Iqbal K, McLachlan DRS, Winblad B, Wisniewsky HM. Alzheimer's
    disease: brain mechanisms, diagnosis and therapeutic strategies.
    New York, John Wiley & Sons Ltd, 1991: 27-33

39. Bruhn P, and the Copenhagen Study Group of Neurological
    Complications in AIDS. AIDS and dementia; a quantitative
    neuropsychological study of unselected Danish patients.
    Acta Neurol Scand 1987; 76: 443-47

40. Bruyn GW.
    Huntington's chorea: a historical, clinical and laboratory synopsis.
    In: Vinken P, Bruyn GW (Eds). Handbook of clinical neurology vol 6.
    Amsterdam, North Holland Publishing Co, 1968: 298-378

41. Butters N, Salmon DP, Heindel W, Granholm E.Episodic, semantic,
    and procedural memory: some comparisons of Alzheimer and
    Huntington disease patients.
    in: Terry RD (Ed). Ageing and the brain.
    New York, Raven Press, 1988: 63-87

42. Butters N, Granholm E, Salmon DP, et al.
    Episodic and semantic memory: a comparison of amnesic and demented
    patients.
    J Clin Exp Neuropsychol 1987; 9: 479-97

43. Butters N, Sax D, Montgomery K, Tarlow S.
Comparison of the neuropsychological deficits associated with early and advanced Huntington's disease.
Arch Neurol 1978; 35: 585-89

44. Caine ED, Bamford KA, Schiffer RB, et al.
A controlled neuropsychological comparison of Huntington's disease and Multiple sclerosis.
Arch Neurol 1986; 43: 249-54

45. Caine ED, Fisher JM.
Dementia in Huntington's disease.
In: Frederiks JAM (ed) Handbook of clinical neurology, vol 2.
Amsterdam, Elsevier Science Publ BV, 1985: 305-10

46. Caine ED.
Pseudodementia: current concepts and future directions.
Arch Gen Psychiatr 1981; 38: 1359-64

47. Callanan MM, Logsdail SJ, Ron MA, Warrington EK.
Cognitive impairment in patients with clinically isolated lesions of the type seen in multiple sclerosis.Brain 1989; 112: 361-74

48. Caltagirone C, Carlesimo A, Nocentini U, Vicari S.
Differential aspects of cognitive impairment in patients suffering from Parkinson's and Alzheimer's disease: a neuropsychological evaluation.
Intern J Neurosci 1989; 44: 1-7

49. Caltagirone C, Carlesimo A, Nocentini U, Vicari S.
Defective concept formation in Parkinsonians is independent from mental deterioration.
J Neurol Neurosurg Psychiatr 1989; 52: 334-37

50. Cambier J, Masson M, Viader F, et al.
Le syndrome frontal de la paralysie supranucléaire progressive.
Rev Neurol (Paris) 1985; 141, 8-9; 528-36

51. Caplan LR.
Binswanger's disease.
In: Frederiks JAM (Ed). Handbook of clinical neurology, vol 2.
Amsterdam, Elsevier Science Publ BV, 1985: 317-21

52. Caplan LR, Schoene WC.
Clinical features of subcortical arteriosclerotic encephalopathy (Binswanger disease).
Neurol 1978; 28: 1206-15

53. Cho Chia Yuen GKK, Schoonderwalt HC.
Neuropsychologische aspecten van normal pressure hydrocephalus: een case presentatie. (Neuropsychological aspects of normal pressure hydrocephalus: a case presentation.)
Tijdschr Gerontol Geriatr 1983; 14: 6-10

54. Chui HC.
Dementia: a review emphasizing clinicopathologic correlation and brain-behavior relationships.
Arch Neurol 1989; 46: 806-14

55. Chui HC, Mortimer JA, Slager U, et al.
Pathological correlates of dementia in Parkinson's disease.
Arch Neurol 1986; 43: 991-95

56. Collignon R, Rectem D, Laterre EC, et al.
Neuropsychological aspects of normal pressure hydrocephalus.
In: Meyer JS, Lechner H, Reivich M (Eds). Cerebral vascular disease.
Amsterdam, Excerpta Medica, 1977: 62-63

57. Croisile B, Laurent B, Michel D et al.
Différentes modalités cliniques des aphasies dégéneratives.
Rev Neurol 1991; 147: 192-199

58. Croisile B, Trillet M, Hibert O et al.
Désordres visuo-constructifs et alexie-agraphie associés à une atrophie corticale postérieure.
Rev Neurol 1991; 147: 138-143

59. Cummings JC, Benson DF.
Dementia: a clinical approach.
Boston, Butterworth, 1983 (2nd ed., 1992)

60. Cummings JL.
The dementias of Parkinson's disease: prevalence, characteristics, neurobiology, and comparison with the dementia of Alzheimer's type.
Eur Neurol 1988; 28 (suppl 1): 15-23

61. Cummings JL, Darkins A, Mendez M.
    Alzheimer's disease and Parkinson's disease: comparison of speech and
    language alterations.
    Neurology 1988; 38: 680-684

62. Cummings JL.
    Subcortical dementia: neuropsychology, neuropsychiatry, and
    pathophysiology.
    Br J Psychiat 1986; 149: 682-97

63. Cummings JL, Benson DF.
    Subcortical dementia: review of an emerging concept.
    Arch Neurol 1984; 41: 874-79

64. Cutler NR, Haxby JV, Duara R, et al.
    Clinical history, brain metabolism, and neuropsychological function in
    Alzheimer's disease.
    Ann Neurol 1985; 18: 298-309

65. D'Antona R, Baron JC, Samson Y, et al.
    Subcortical dementia: frontal cortex hypometabolism detected by
    positron emission tomography in patients with progressive
    supranuclear palsy.
    Brain 1985; 108: 785-99

66. Dal Canto M.
    AIDS and the nervous system: current status and future perspectives.
    Hum Pathol 1989; 20: 410-18

67. Damasio AR.
    The frontal lobes.
    In: Heilman KM, Valenstein E (Eds). Clinical neuropsychology, 2nd ed.
    New York, Oxford University Press, 1985: 339-75

68. De Gans J, Portegies P, Derix MMA, et al.
    Het AIDS-Dementiecomplex: een primaire infectie met humaan
    immunodeficientievirus type I. (The AIDS Dementia Complex: a
    primary infection with human immunodeficiency virus type 1.)
    Ned Tijdschr Geneeskd 1988; 132: 1570-75

69. De La Monte S, Well SE, Hedley-White ET, Growdon JH.
    Neuropathological distinction between Parkinson's dementia and
    Parkinson's plus Alzheimer's disease.
    Ann Neurol 1989; 26: 309-320

70. De la Monte SM, Von Sattel JP, Richardson EP.
Morphometric demonstration of atrophic changes in the cerebral
cortex, white matter, and neostriatum in Huntington's disease.
J Neuropathol Exp Neurol 1988; 47: 516-25

71. DeLong MR.
Primate models of movement disorders of basal ganglia origin.
Trends Neurosci 1990; 13(7): 281-85

72. Derix MMA, de Gans J, Stam J, Portegies P.
Mental changes in patients with AIDS.
Clin Neurol Neurosurg 1990; 92: 215-22

73. Derix MMA, Groet E.
Subcorticale en corticale dementie: een neuropsychologisch
onderscheid. (Subcortical and cortical dementia: a neuropsychological
distinction.)
In: Schroots JJE, Bouma A, Braam GPA, et al. (Eds). Gezond
zijn is ouder worden.
Assen, van Gorcum, 1989: 179-88

74. Derix MMA, Hijdra A, Verbeeten BWJ. Mental changes in subcortical
arteriosclerotic encephalopathy.
Clin Neurol Neurosurg 1987; 89: 71-8

75. Di Rocco C, Di Trapani G, Maira C, et al.
Anatomo-clinical correlations in normotensive hydrocephalus.
J Neurol Sci 1977:; 33: 437-52

76. DSM-III-R. Diagnostic and Statistical Manual of Mental Disorders, ed.
3, revised.
Washington D.C, American Psychiatric Association, 1987

77. Dubois B, Pillon B, Legault F, et al.
Slowing of cognitive processing in progressive supranuclear palsy.
Arch Neurol 1988; 45: 1194-99

78. Ebers GC.
Multiple sclerosis and other demyelinating diseases.
In: Ashbury AK, McKhann GM, McDonald WI (Eds). Diseases of the
nervous system, vol 2.
London, W Heineman Med Books, 1986: 1268-81

79. El-Awar M, Becker JT, Hammond KM, et al.
    Learning deficit in Parkinson's disease: comparison with Alzheimer's
    disease and normal aging.
    Arch Neurol 1987; 44: 180-4

80. Ellis AW, Young AW.
    Human cognitive neuropsychology.
    London, Lawrence Erlbaum Associates Publ, 1988

81. Fahn S.
    Parkinson's disease and other basal ganglia disorders.
    In: Asbury AK, McKahnn GM, McDonald WI (Eds). Diseases of the
    nervous system vol 2.
    London, W Heineman Med Books, 1986: 1217-28

82. Filley CM, Franklin GM, Heaton RK, Rosenberg NL.
    White matter dementia; clinical disorders and implications.
    Neuropsychiatr Neuropsychol Behav Neurol, 1989; 1: 239-54

83. Fischer PA, Enzensberger W.
    Neurological complications in AIDS.
    J Neurol 1987; 234: 269-79

84. Folstein SE, Leigh RJ, Parhad IM, Folstein MF.
    The diagnosis of Huntington's disease.
    Neurol 1986; 36: 1279-83

85. Franklin GM, Nelson LM, Filley CM, Heaton RK.
    Cognitive loss in multiple sclerosis; case reports and review of the
    literature.
    Arch Neurol 1989; 46: 162-67

86. Freedman M, Oscar-Berman M.
    Tactile discrimination learning deficits in Alzheimer's and Parkinson's
    disease.
    Arch Neurol 1987; 44: 394-398

87. Freedman M, Oscar-Berman M.
    Comparative neuropsychology of cortical and subcortical dementia.
    Can J Neurol Sci 1986; 13: 410-14

88. Freedman M, Oscar-Berman M.Selective delayed response deficits in
    Parkinson's disease and Alzheimer's disease.
    Arch Neurol 1986; 43: 886-90

89. Freedman M, Albert ML.
    Subcortical dementia.
    In: Frederiks JAM (Ed). Handbook of Clinical Neurology, vol 2.
    Amsterdam, Elsevier Science Publ BV, 1985: 311-15

90. Friedland RP, Jagust WJ, Huesman RH, et al.
    Regional cerebral glucose transport and utilization in Alzheimer's
    disease.
    Neurol 1989; 39: 1427-34

91. Funnell E.
    Modelling cognitive function in dementia.
    In: Stahl SM, Iversen SD, Goodman EC (Eds). Cognitive
    neurochemistry.
    Oxford, Oxford University Press, 1987: 37-45

92. Fuster JM.
    The prefrontal cortex. Anatomy, physiology, and neuropsychology of
    the frontal lobes. 2nd ed.
    New York, Raven Press, 1989

93. Gabuzda DA, Hirsch MS.
    Neurologic manifestations of infections with human immunodeficiency
    virus.
    Ann Int Med 1987: 107:383-91

94. Girotti F, Marano R, Soliveri P, et al.
    Relationship between motor and cognitive disorders in Huntington's
    disease.
    J Neurol 1988; 235: 454-57

95. Goffinet AM, De Volder AG, Gillain C, et al.
    Positron tomography demonstrates frontal lobe hypometabolism in
    progressive supranuclear palsy.
    Ann Neurol 1989; 25: 131-39

96. Goldberg TE, Berman KF, Mohr E, Weinberger DR.
    Regional cerebral blood flow and cognitive function in Huntington's
    disease and schizophrenia: a comparison of patients matched for
    performance on a prefrontal-type task.
    Arch Neurol 1990; 47: 418-22

97. Goldenberg G, Wimmer A, Auff E, Schnabert G.
     Impairment of motor planning in patients with Parkinson's disease:
     evidence from ideomotor apraxia testing.
     J Neurol Neurosurg Psychiatr 1986; 49: 1266-72

98. Goldman-Rakic PS.
     Circuit basis of a cognitive function in non-human primates
     In: Stahl SM, Iversen SD, Goodman EC (Eds). Cognitive
     neurochemistry.
     Oxford, Oxford University Press, 1987: 90-110

99. Goodin DS, Aminoff MJ.
     Electrophysiological differences between subgroups of dementia.
     Brain 1986; 109: 1103-13

100. Gotham AM, Brown RG, Marsden CD.
     'Frontal' cognitive function in patients with Parkinson's disease 'on'
     and 'off' levodopa.
     Brain 1988; 111: 299-321

101. Graff-Radford NR, Rezal K, Godersky JC, et al.
     Regional cerebral blood flow in normal pressure hydrocephalus. J
     Neurol Neurosurg Psychiatr 1987; 50: 1589-96

102. Graff-Radford NR, Godersky J.
     Normal pressure hydrocephalus. Onset of gait abnormality before
     dementia predicts a good surgical outcome.
     Arch Neurol 1986; 43: 940-42

103. Grafman J, Litvan J, Gomez C, Chase TN.
     Frontal lobe function in progressive supranuclear palsy.
     Arch Neurol 1990; 47: 553-58

104. Grant I, Atkinson JH, Hesslink JR, et al.
     Evidence for early central nervous system involvement in the acquired
     immunodeficiency syndrome (AIDS) and other human
     immunodeficiency virus (HIV) infections.
     Ann Intern Med 1987; 107: 828-36

105. Grant I.
     Neuropsychological and psychiatric disturbances in multiple sclerosis.
     In: McDonald WI, Silberberg DH (Eds). Multiple sclerosis.
     London, Butterworth, 1986: 134-52

106. Grant I, McDonald WJ, Trimble MR, et al.
Deficient learning and memory in early and middle phases of multiple sclerosis.
J Neurol Neurosurg Psychiatr 1984; 47: 250-55

107. Gray F, Gherardi R, Scaravilli F.
The neuropathology of the acquired immune deficiency syndrome (AIDS).
Brain 1988; 111: 245-66

108. Green J, Morris JC, Sandson J et al. Progressive aphasia: a precursor of global dementia?
Neurology 1990; 40: 423-29

109. Gustafson L.
Dementia, cognitive impairment and learning difficulties.
Current Opinion in Neurology and Neurosurgery 1991; 4: 91-7

110. Gustafson L, Hagberg B.
Recovery in hydrocephalic dementia after shunt operation.
J Neurol Neurosurg Psychiatr 1978; 41: 940-47

111. Harrington DL, Haaland KY, Yeo RA, Marder E.
Procedural memory in Parkinson's disease: impaired motor but not visuoperceptual learning.
J Clin Exp Neuropsychol 1990; 12(2): 323-39

112. Heaton R, Nelson LM, THompson DS, et al.
Neuropsychological findings in relapsing-remitting and chronic-progressive multiple sclerosis.
J Consult Clin Psychol 1985; 53: 103-10

113. Heindel WC, Salmon DP, Shults CW, et al.
Neuropsychological evidence for multiple implicit memory systems: a comparison of Alzheimer's, Huntington's, and Parkinson's disease patients.
The Journal of Neuroscience, 1989; 9(2): 582-87

114. Helkala E-L, Laulumaa V, Soininen H, Riekkinen PJ.
Recall and recognition memory in patients with Alzheimer's and Parkinson's disease.
Ann Neurol, 1988; 24: 214-17

115. Hier DB, Thomas C, Shindler AG.
A case of subcortical dementia due to sarcoidosis of the hypothalamus
and fornices.
Brain and Cognition 1983; 2: 189-98

116. Hietanen M, Teräväinen H.
The effect of age of disease onset on neuropsychological performance in
Parkinson's disease.
J Neurol Neurosurg Psychiatr 1988; 51: 244-49

117. Hietanen M, Teräväinen H.
Cognitive performance in early Parkinson's disease.
Acta Neurol Scand, 1986; 73: 151-59

118. Hijdra A.
Vascular dementia.
In: Bradley WG, Daroff RB, Fenichel GM, Marsden CD (Eds).
Neurology in clinical practice.
Boston, Butterworth, 1990: 1425-35

119. Hoffman JM, Guze BH, Baxter LR, et al.
(^{18}F)-Fluorodeoxyglucose (FDG) and positron emission tomography
(PET) in aging and dementia.
Eur Neurol 1989; 29 (suppl 3): 16-24

120. Holland JC, Tross S.
The psychosocial and neuropsychiatric sequelae of the acquired
immunodeficiency syndrome and related disorders.
Ann Int Med 1985; 103: 760-64

121. Huber SJ, Shuttleworth EC, Christy JA, et al.
Magnetic resonance imaging in dementia of Parkinson's disease. J
Neurol Neurosurg Psychiatr 1989; 52: 1221-27

122. Huber SJ, Shuttleworth EC, Freidenberg DL.
Neuropsychological differences between the dementias of Alzheimer's
and Parkinson's disease.
Arch Neurol, 1989; 46: 1287-91

123. Huber SJ, Paulson GW, Shuttleworth EC.
Relationship of motor symptoms, intellectual impairment, and
depression in Parkinson's disease.
J Neurol Neurosurg Psychiatr 1988; 51: 855-58

124. Huber SJ, Paulson GW, Shuttleworth EC, et al.
Magnetic resonance imaging correlates of dementia in multiple sclerosis.
Arch Neurol 1987; 44: 732-36

125. Huber SJ, Shuttleworth EC, Paulson GW.
Dementia in Parkinson's disease.
Arch Neurol 1986; 43: 987-90

126. Huber SJ, Shuttleworth EC, Paulson GW, et al.
Cortical versus subcortical dementia: neuropsychological differences.
Arch Neurol 1986; 43: 392-94

127. Ingram RE, Kendall PC.
Cognitive clinical psychology: implications of an information processing perspective.
In: Ingram RE (Ed). Information processing approaches to clinical psychology.
New York, Academic Press Inc, 1986: 6-21

128. Jacobs D, Tröster AI, Butters N, et al. Intrusion errors on the visual reproduction tests of the Wechsler Memory Scale and the Wechsler Memory Scale-Revised: an analysis of demented and amnesic patients.
The Clin Neuropsychologist 1990; 4(2): 177-91

129. Jagust WJ, Friedland RP, Budinger TF.
Positron emission tomography with 18F-Fluorodeoxyglucose differentiates normal pressure hydrocephalus from Alzheimer-type dementia.
J Neurol Neurosurg Psychiatr 1985; 48: 1091-96

130. Janati A, Appel AR.
Psychiatric aspects of progressive supranuclear palsy.
J Nerv Ment Dis, 1984; 172(2): 85-89

131. Janota I.
Dementia, deep white matter damage and hypertension: 'Binswanger's disease'.
Psychol Med 1981; 11: 39-48

132. Janssen RS, Saykin AJ, Cannon L, et al.
Neurological and neuropsychological manifestations of HIV-1 infection: association with AIDS-related complex but not asymptomatic HIV-1 infection.
Ann Neurol 1989; 26: 592-600

133. Janssen RS, Sayking AJ, Kaplan JE, et al.
Neurological complications of human immunodeficiency virus infection
in patients with lymphadenopathy syndrome.
Ann Neurol 1988; 23: 49-55

134. Jason GW, Pajurkova EM, Suchowerksy O, et al.
Presymptomatic neuropsychological impairment in Huntington's
disease.
Arch Neurol 1988; 45: 769-73

135. Jellinger K.
Overview of morphological changes in Parkinson's disease.
Adv in Neurol 1986; 45:1-18

136. Jennekens-Schinkel A, Laboyrie PM, Lanser JBK, Van der Velde EA.
Cognition in patients with multiple sclerosis.
J Neurol Sci 1990; 99: 229-47

137. Jennekens-Schinkel A, Lanser JBK, Van der Velde EA, Sanders
EACM.
Performances of multiple sclerosis patients in tasks requiring language
and visuoconstruction: assessment of outpatients in quiescent disease
stages.
J Neurol Sci 1990; 95: 89-103

138. Jennekens-Schinkel A, Van der Velde AE, Sanders EACM, Lanser
JBK.
Memory and learning in outpatients with quiescent multiple sclerosis.
J Neurol Sci 1990; 95: 311-25

139. Jennekens-Schinkel, Van der Velde EA, Sanders EACM, Lanser JBK.
Visuospatial problem solving, conceptual reasoning and sorting
behaviour in multiple sclerosis out-patients.
J Neurol Sci 1989; 90: 187-201

140. Jennekens-Schinkel A, Sanders EACM, Lanser JBK, Van der Velde EA.
Reaction time in ambulant multiple sclerosis patients; part I. Influence
of prolonged cognitive effort.
J Neurol Sci 1988; 85: 173-86

141. Jennekens-Schinkel A, Sanders EACM, Lanser JBK, Van der Velde EA.
Reaction time in ambulant multiple sclerosis patients; part II. Influence of task complexity.
J Neurol Sci 1988; 85: 187-96

142. Jennekens-Schinkel A, Sanders EACM.
Decline of cognition in multiple sclerosis: dissociable deficits.
J Neurol Neurosurg Psychiatr 1986; 49: 1354-60

143. Jernigan TL, Butters N.
Neuropsychological and neuroradiological distinctions between Alzheimer's and Huntington's diseases.
Neuropsychology 1989; 3: 283-90

144. Josiassen RC, Curry LM, Manchall EL.
Development of neuropsychological deficits in Huntington's disease.
Arch Neurol 1983; 40: 791-76

145. Josiassen RC, Curry L, Roemer RA, DeBease C.
Patterns of intellectual deficit in Huntington's disease.
J Clin Neuropsychol 1982; 4: 173-83

146. Junqué C, Pujol J, Vendrell P, et al.
Leuko-araiosis on magnetic resonance imaging and speed of mental processing.
Arch Neurol 1990; 47: 151-56

147. Kaemingk KL, Kasniak AW.
Neuropsychological aspects of human immunodeficiency virus infection.
The Clin Neuropsychologist 1989; 3: 309-26

148. Kapur N.
Memory disorders in clinical practice.
London, Butterworth, 1988

149. Katzman R.
Normal pressure hydrocephalus.
In: Wells CE (Ed). Dementia. 2nd ed.
Philadelphia, FA Davis, 1977: 69-92

150. Kelly MP, Kirshner HS.
Syndromes of the frontal lobes.
In: Kirshner HS. Behavioral neurology.
New York, Churchill Livingstone, 1986: 101-20

151. Kinkel WR, Jacobs L, Polachine I, et al.
Subcortical arteriosclerotic encephalopathy (Binswanger's disease):
computed tomographic, nuclear magnetic resonance, and clinical
correlations.
Arch Neurol 1985; 42: 951-59

152. Kirshner HS.
Behavioral neurology, chapter 5. Apraxias.
New York, Churchill Livingstone, 1986: 59-68

153. Kirshner HS, Webb WG, Kelly MP, Wells ChE.
Language disturbance; an initial symptom of cortical degenerations and
dementia.
Arch Neurol 1984; 41: 491-496

154. Knopman DS, Mastri AR, Frey WH et al.
Dementia lacking distinctive histologic features; a common non-
Alzheimer degenerative dementia.
Neurology 1990; 40: 252-256

155. Knopman DS, Mastri AR, Frey WH, et al.
The spectrum of imaging and neuropsychological findings in Pick's
disease.
Neurology 1989; 39: 362-368

156. Kovner R, Perecman E, Lazar W, et al.
Relation of personality and attentional factors to cognitive deficits in
human immunodeficiency virus-infected subjects.
Arch Neurol 1989; 46: 274-77

157. Kristensen MO.
Progressive supranuclear palsy - 20 years later.
Acta Neurol Scand 1985; 71: 177-89

158. Kuhl DE, Phelps ME, Markham CH, et al.
Cerebral metabolism and atrophy in Huntington's disease determined
by 18FDG and computed tomographic scan.
Ann Neurol 1982; 12: 425-34

159. Kuwert T, Lange HW, Langen K-J, et al.
Cortical and subcortical glucose consumption measured by PET in patients with Huntington's disease.
Brain 1990; 113: 1405-23

160. Lavernhe G, Pollack P, Brenier F, et al.
Maladie d'Alzheimer et maladie de Parkinson. Différenciation neuropsychologique.
Rev Neurol (Paris) 1989; 145: 24-30

161. Lee A, Yu YL, Tsoi M, et al.
Subcortical arteriosclerotic encephalopathy - a controlled psychometric study.
Clin Neurol Neurosurg 1989; 91(3): 235-41

162. Leenders KL, Frackowiak RSJ, Lees AJ.
Progressive supranuclear palsy: brain energy metabolism, blood flow and fluorodopa uptake measured by positron emission tomography.
Brain 1988; 111: 615-30

163. Lees AJ.
The Steele-Richardson-Olszweski syndrome (progressive supranuclear palsy).
In: Marsden CD, Fahn S (Eds). Movement disorders 2.
London, Butterworth, 1987: 272-87

164. Lees AJ, Smith AJ.
Cognitive deficits in the early stages of Parkinson's disease.
Brain 1983; 106: 257-70

165. Levin BE, Llabre MM, Weiner WJ.
Cognitive impairments associated with early Parkinson's disease.
Neurol 1989; 39: 557-61

166. Lezak MD.
Neuropsychological assessment, 2nd ed. (3rd ed, 1994 in press)
New York, Oxford University Press, 1983

167. Litvan I, Grafman J, Gomez C, Chase TN.
Memory impairment in patients with progressive supranuclear palsy.
Arch Neurol 1989; 46: 765-67

168. Litvan I, Grafman J, Vendrell P, Martinez JM.
Slowed information processing in multiple sclerosis.
Arch Neurol 1988; 45: 281-85

169. Litvan I, Grafman J, Vendrell P, et al.
Multiple memory deficits in patients with multiple sclerosis.
Arch Neurol 1988; 45: 607-10

170. Loeb C.
The lacunar syndromes.
Eur Neurol 1989; 29 (suppl): 2-7

171. Loeb C.
Vascular dementias.
In: Frederiks JAM (Ed). Handbook of clinical neurology, vol 2.
Amsterdam, Elsevier Science Publ BV, 1985: 353-69

172. Loizou LA, Kendall BE, Marshall J.
Subcortical arteriosclerotic encephalopathy; a clinical and radiological
investigation.
J Neurol Neurosurg Psychiatr 1981; 44: 294-304

173. Loranger AW, Goodell H, McDowell F.
Intellectual impairment in Parkinson's syndrome.
Brain 1972; 95: 402-12

174. Luria AR.
The working brain.
New York, Penguin Books Ltd, 1978

175. Lyon-Caen O, Jouvent R, Hauser S, et al.
Cognitive function in recent-onset demyelinating diseases.
Arch Neurol 1986; 43: 1138-41

176. MacArthur JC, Becker PS, Parisi JE, et al.
Neuropathological changes in early HIV-1 dementia.
Ann Neurol 1989; 26: 681-84

177. Maher ER, Lees EAJ.
The clinical features and natural history of the Steele Richardson
Olszewski syndrome (progressive supranuclear palsy).
Neurol 1986; 36: 1005-08

178. Maher ER, Smith EM, Lees AJ.
Cognitive deficits in the Steele Richardson Olzewski syndrome
(progressive supranuclear palsy).
J Neurol Neurosurg Psychiatr 1985; 48: 1234-39

179. Martin A, Brouwers P, Lalonde F et al.
Towards a behavioral typology of Alzheimer's patients.
J Clin Exp Neuropsychol 1986; 8: 594-610

180. Martin JB, Gusella JF.
Huntington's disease: pathogenesis and management.
N Eng J Med 1986; 315: 1267-86

181. Martone M, Butters N, Payne M, et al.
Dissociations between skill learning and verbal recognition in amnesia
and dementia.
Arch Neurol 1984; 41: 965-70

182. Matthews WB.
Clinical Aspects.
In: Matthews WB (Ed). McAlpine's Multiple Sclerosis.
Edingburgh, Churchill Livingstone, 1985: 49-280

183. Mayeux R, Stern Y, Sano M, et al.
Clinical and biochemical correlates of bradyphrenia in Parkinson's
disease.
Neurol 1987; 37: 1130-34

184. Mayeux R, Stern Y, Herman A, et al.
Correlates of early disability in Huntington's disease.
Ann Neurol 1986; 20: 727-31

185. Mayeux R, Stern Y, Spanton S.
Heterogeneity in dementia of the Alzheimer type: evidence of subgroups.
Neurology 1985; 35: 453-61

186. Mayeux R, Stern Y, Rosen J, Benson DF.
Is "subcortical dementia" a recognizable clinical entity?.
Ann Neurol 1983; 14: 278-83

187. Mayeux R.
Depression and dementia in Parkinson's disease.
In: Marsden CD, Fahn S (Eds). Neurology vol 2.
London, Butterworth, 1982: 75-95

188. Mayeux R, Stern Y, Rosen J, Leventhal L.
Depression, intellectual impairment and Parkinson's disease.
Neurol 1981; 31: 645-50

189. McArthur JC, Becker PS, Parisi JE, et al.
Neuropathological changes in early HIV-1 Dementia.
Ann Neurol 1989; 26: 681-84

190. McHugh PR, Folstein MF.
Psychiatric syndromes of Huntington's chorea: a clinical and
phenomenologic study.
In: Benson DF, Blumer D (Eds). Psychiatric aspects of neurological
diseases.
New York, Grune & Stratton, 1975: 267-85

191. McHugh PR.
Occult hydrocephalus.
Q J Med. 1964; 33: 297-308

192. Medaer R, Nelissen E, Appel B, et al.
Magnetic resonance imaging and cognitive functioning in multiple
sclerosis.
J Neurol 1987; 235: 86-89

193. Mehler MF, Dickson D, Davies P, Horoupian DS.
Primary dysphasic dementia: clinical, pathological, and biochemical
studies.
Neurology 1986; 20: 126

194. Mesulam M-M.
Patterns in behavioral neuroanatomy: association areas, the limbic
system, and hemispheric specialization.
In: Mesulam M-M (Ed). Principles of behavioral neurology.
Philadelphia, FA Davis Company, 1985: 1-70

195. Mesulam M-M.
Attention, confusional states, and neglect.
In: Mesulam M-M (Ed). Principles of behavioral neurology.
Philadelphia, FA Davis Company, 1985: 125-68

196. Milberg W, Albert M.
Cognitive differences between patients with progressive supranuclear
palsy and Alzheimer's disease.
J Clin Exp Neuropsychol 1989; 11: 605-14

197. Millers EN, Selnes OA, McArthur JC, et al.
Neuropsychological performance in HIV-1-infected homosexual men:
the multicenter AIDS cohort study (MACS).
Neurol 1990; 40: 197-203

198. Mohr E, Fabrini G, Ruggieri S, et al.
Cognitive concomitants of dopamine system stimulation in
Parkinsonian patients.
J Neurol Neurosurg Psychiatr 1987; 50: 1192-96

199. Morris JC, Cole M, Banker BQ, Wright D.
Hereditary dysphasic dementia and the Pick-Alzheimer spectrum.
Ann Neurol 1984; 16: 455-66

200. Morris RG, Downes JJ, Sahakian BJ, et al.
Planning and spatial working memory in Parkinson's disease.
J Neurol Neurosurg Psychiatr 1988; 51: 757-66

201. Mortimer JA, Christensen KJ, Webster D.
Parkinson dementia.
In: Frederiks JAM (Ed). Handbook of clinical neurology, vol 2.
Amsterdam, Elsevier Science Publ BV, 1985: 371-84

202. Moss BM, ALbert MS, Butters N, Payne M.
Differential patterns of memory loss among patients with Alzheimer's
disease, Huntington's disease, and alcoholic Korsakoff syndrome.
Arch Neurol 1986; 43: 239-46

203. Navia BA, Price RW.
The acquired immunodeficiency syndrome dementia complex as the
presenting or sole manifestation of human immunodeficiency virus
infection.
Arch Neurol 1987; 44: 65-69

204. Navia BA, Cho E-S, Petito CK, Price RW.
The AIDS Dementia Complex: II. Neuropathology.
Ann Neurol 1986; 19: 525-35

205. Navia BA, Jordan BD, Price RW.
The AIDS Dementia Complex: I: Clinical features.
Ann Neurol 1986; 517-24

206. Navia BA, Petito CK, Gold JWM, et al.
Cerebral toxoplasmosis complicating the acquired immune deficiency
syndrome: clinical and neuropathological findings in 27 patients.
Ann Neurol 1986; 19: 224-38

207. Neary D, Snowden JS, Northen B, Goulding P.
Dementia of the frontal lobe type.
J Neurol Neurosurg Psychiatr 1988; 51; 353-361

208. Neary D, Snowden JS, Shields RA et al.
Single photon emission tomography using ^{99m}Tc-HM-PAO in the
investigation of dementia.
J Neurol Neurosurg Psychiatr 1987; 50: 1101-1109

209. Neary D, Snowden JS, Bowen DM, Sims NR et al.
Neuropsychological syndromes in presenile dementia due to cerebral
atrophy.
J Neurol Neurosurg Psychiatr 1986; 49; 163-74

210. Nieuwenhuys R, Voogd J, van Huijzen Chr.
The human central nervous system: a synopsis and atlas.
Berlin, Springer Verlag, 1978

211. Norman DA, Shallice T.
Attention to action: Willed and automatic control of behaviour.
In: Davidson RJ, Schwartz GE, Shapiro D (eds). Consciousness and
self-regulation: Advances in research and theory, vol 4.
New York, Plenum Press, 1986: 1-18

212. Ojemann RG, Fisher CM, Adams RD, et al.
Further experience with the syndrome of "normal" pressure
hydrocephalus.
J Neurosurg 1969; 31: 279-94

213. Parent A.
Extrinsic connections of the basal ganglia.
Trends Neurosci 1990; 13(7): 254-58

214. Perdices M, Cooper DA.
Neuropsychological investigation of patients with AIDS and ARC.
J Acquired Immune Deficiency Syndrome 1990; 3: 555-64

215. Perdices M, Cooper DA.
Simple and choice reaction time in patients with human
immunodeficiency virus infection.
Ann Neurol 1989; 25: 460-67

216. Perlmutter JS.
New insights into the pathophysiology of Parkinson's disease: the
challenge of positron emission tomography.
Trends Neurosci 1988; 11: 203-08

217. Peyser JM, Edwards KR, Poser CM, Filskov SB.
Cognitive function in patients with multiple sclerosis.
Arch Neurol 1980; 37: 577-79

218. Piccirilli M, D'Allessandro P, Finali G, et al.
Frontal lobe dysfunction in Parkinson's disease: prognostic value for
dementia?
Eur Neurol 1989; 29: 71-76

219. Pillon B, Dubois B, Ploska A, Agid Y.
Severity and specificity of cognitive impairment in Alzheimer's, Hunting-
ton's, and Parkinson's diseases and progressive supranuclear palsy.
Neurology 1991; 41: 634-43

220. Pillon B, Dubois B, Bonnet A-M, et al.
Cognitive slowing in Parkinson's disease fails to respond to levodopa
treatment.
Neurol 1989; 39: 762-68

221. Pillon B, Dubois B, Cusimano G, et al.
Does cognitive impairment in Parkinson's disease result from non-
dopaminergic lesions?
J Neurol Neurosurg Psychiatr 1989; 52: 201-06

222. Pillon B, Dubois B, L'Hermitte F, Agid Y.
Heterogeneity of intellectual impairment in progressive supranuclear
palsy, Parkinson's disease and Alzheimer's disease.
Neurol 1986; 36: 1179-85

223. Poeck K, Luzatti C.
Slowly progressive aphasia in three patients.
Brain 1988; 111: 151-168

224. Portegies P, Epstein L, Tjong A, et al. Human immunodeficiency virus
type 1 antigen in cerebrospinal fluid. Correlation with clinical
neurological status.
Arch Neurol 1989; 46: 261-64

225. Price RW, Brew B, Sidtis J, et al.
The brain in AIDS: Central nervous system HIV-1 infection and AIDS
Dementia Complex.
Science 1988; 239: 586-92

226. Price RW, Navia BA, Pumarola-Sune T, et al.
The Aids Dementia Complex: answers and questions.
In: Gluckman JC, Vilmer E (Eds). Acquired Immunodeficiency
Syndrome.
Paris, Elsevier, 1987: 205-10

227. Rafal RD, Posner MI, Friedman JH, et al.
Orienting of visual attention in progressive supranuclear palsy.
Brain 1988; 111: 267-80

228. Ransmayer G, Schmidhuber-Eiler B, Karamat E, et al.
Visuoperception and visuospatial and visuorotational performance in
Parkinson's disease.
J Neurol 1987; 235: 99-101

229. Rao SM, Leo GJ, Haughton VM, et al.
Correlation of magnetic resonance imaging with neuropsychological
testing in multiple sclerosis.
Neurol 1989; 39: 161-66

230. Rao SM, St. Aubin-Faubert P, Leo GJ.
Information processing speed in patients with multiple sclerosis.
J Clin Exp Neuropsychol 1989; 11(4): 471-77

231. Rao SM.
Neuropsychology of multiple sclerosis: a critical review.
J Clin Exp Neuropsychol 1986; 8: 503-43

232. Rao SM, Hammeke TA, McQuillen MP, et al.
Memory disturbance in chronic progressive multiple sclerosis.
Arch Neurol 1984; 41: 625-31

233. Révész T, Hawkins CP, du Boulay EPGH, et al.
Pathological findings correlated with magnetic resonance imaging in subcortical arteriosclerotic encephalopathy (Binswanger's disease).
J Neurol Neurosurg Psychiatr 1989; 52: 1337-44

234. Rinne JO, Rummukainen J, Paljärvi L, Rinne UK.
Dementia in Parkinson's disease is related to neuronal loss in the medial substantia nigra.
Ann Neurol 1989; 26: 47-50

235. Rinne UK, Laakso K, Mölsä P, et al.
Dementia and brain receptor changes in Parkinson's disease and in senile dementia of the Alzheimer type.
In: Gottfries CG (Ed). Normal aging, Alzheimer's disease and senile dementia: aspects on etiology, pathogenesis, diagnosis and treatment.
Bruxelles, Editions de l'Université de Bruxelles, 1985: 135-46

236. Robbins TW, Everitt BJ.
Psychopharmacological studies of arousal and attention.
In: Stahl SM, Iversen SD, Goodman EC (Eds). Cognitive neurochemistry.
Oxford, Oxford University Press, 1987: 135-70

237. Rogers D, Lees AJ, Smith E, et al. Bradyphrenia in Parkinson's disease and psychomotor retardation in depressive illness.
Brain 1987; 110: 761-76

238. Román GC.
Senile dementia of the Binswanger type: a vascular form of dementia in the elderly.
JAMA 1987; 258: 1782-88

239. Rossor M.
The neurochemistry of cortical dementias.
In: Stahl SM, Iversen SD, Goodman EC (Eds). Cognitive neurochemistry.
Oxford, Oxford University Press, 1987: 233-47

240. Rottenberg DA, Moeller JR, Strother SC, et al.
The metabolic pathology of the AIDS Dementia Complex.
Ann Neurol 1987; 22: 700-6

241. Rubinov DR, Berrehine CH, Brouwers P, Lane HC.
Neuropsychiatric consequences of AIDS.
Ann Neurol 1988; 23 (suppl): S24-26

242. Sacquegna T, Guttman S, Guiliani S, et al.
Binswanger's disease: a review of the literature and a personal
contribution.
Eur Neurol 1989; 29 (suppl): 20-22

243. Sagar HJ, Sullivan EV, Gabrieli JDE, et al.
Temporal ordering and short-term memory deficits in Parkinson's
disease.
Brain 1988; 111: 525-39

244. Sagar HJ, Cohen NJ, Sullivan EV, et al. Remote memory function in
Alzheimer's disease and Parkinson's disease.
Brain 1988; 111: 185-206

245. Sahakian BJ, Morris RG, Evenden JL, et al.
A comparative study of visuospatial memory and learning in Alzheimer-
type dementia and Parkinson's disease.
Brain 1988; 111: 695-718

246. Saint-Cyr JA, Taylor AE, Lang AE.
Procedural learning and neostriatal dysfunction in man.
Brain 1988; 111: 941-59

247. Salmon DP, Kwo-on-Yuen PR, Heindel WC, et al.
Differentiation of Alzheimer's disease and Huntington's disease with the
dementia rating scale.
Arch Neurol 1989; 46: 1204-08

248. Salmon JH, Gonen JY, Brown L.
Ventriculoatrial shunt for hydrocephalus ex-vacuo; psychological and
clinical evaluation.
Dis Nerv Sys 1971; 32: 299-307

249. Salmon JH, Armitage JL.
Surgical treatment of hydrocephalus ex-vacuo: ventriculoatrial shunt
for degenerative brain disease.
Neurol 1968; 18: 1123-26

250. Sandson J, Albert ML.
     Perseveration in behavioral neurology.
     Neurol 1987; 37: 1736-41

251. Schmitt FA, Bigley JW, McKinnis R, et al.Neuropsychological outcome
     of zidovudine (AZT) treatment of patients with AIDS and AIDS-related
     complex.
     N Eng J Med 1988; 319: 1573-78

252. Shallice T.
     From neuropsychology to mental structure.
     Cambridge, Cambridge University Press, 1988

253. Shallice T.
     Specific impairments of planning.
     Phil Trans R Soc Lond.B 1982; 298: 199-209

254. Shoulson I.
     Huntington's disease.
     In: Asbury AK, McKahn GM, McDonald WI (Eds). Diseases of the
     nervous system vol 2.
     London, W Heineman Med Books 1986: 1258-67

255. Speelman J.D.
     Parkinson's disease and stereotactic surgery. Thesis
     Amsterdam, University of Amsterdam, 1991

256. Starkenstein SE, Preziosi ThJ, Berthier ML, et al.
     Depression and cognitive impairment in Parkinson's disease.
     Brain 1989; 112: 1141-53

257. Steele JC, Richardson JC, Olszewski J.
     Progressive supranuclear palsy.
     Arch Neurol 1964; 10: 333-59

258. Strub RL, Black FW.
     Neurobehavioral disorders: a clinical approach.
     Philadelphia, FA Davis Company, 1988

259. Strub RL, Black FW.
     Organic brain syndromes.
     Philadelphia, FA Davis, 1981

260. Sullivan EV, Sagar HJ.
Nonverbal recognition and recency discrimination deficits in
Parkinson's disease and Alzheimer's disease.
Brain 1989; 112: 1503-17

261. Sullivan EV, Sagar HJ, Gabrieli JDE, et al.
Different cognitive profiles on standard behavioral tests in Parkinson's
disease and Alzheimer's disease.
J Clin Exp Neuropsychol, 1989; 11(6): 799-820

262. Sypert GW, Leffman H, Ojemann GA.
Occult normal pressure hydrocephalus manifested by parkinsonism-
dementia complex.
Neurol 1973; 23: 234-38

263. Tanahashi N, Meyer JS, Ishikawa Y, et al.
Cerebral blood flow and cognitive testing correlate in Huntington's disease.
Arch Neurol 1985; 42: 1169-75

264. Taylor AE, Saint-Cyr JA, Lang AE.
Parkinson's disease: cognitive changes in relation to treatment
response.
Brain 1987; 110: 35-51

265. Taylor AE, Saint-Cyr JA, Lang AE.
Frontal lobe dysfunction in Parkinson's disease.
Brain 1986; 109: 845-83

266. Thomson AM, Borgesen SE, Bruhn P, Gjerris F.
Prognosis of dementia in normal-pressure hydrocephalus after a shunt
operation.
Ann Neurol 1986; 20: 304-10

267. Tissot R, Constantinides J, Richard J.
Pick's disease.
In: Pick's disease. Frederiks JAM (Ed). Handbook of clinical
neurology, vol. 2.
Amsterdam, Elsevier Science Publ. 1985: 233-245

268. Tross S, Price RW, Navia B, et al.
Neuropsychological characterisation of the AIDS dementia complex: a
preliminary report.
AIDS 1988; 2: 81-88

269. Truelle JL, Palisson E, Le Gall D, et al.
Troubles intellectuels et thymiques dans la sclérose en plaques.
Rev Neurol (Paris) 1987; 143, 8-9: 595-601

270. Tyrell P.J, Sawle GV, Ibanez V, et al.
Clinical and positron emission tomographic studies in the
'extrapyramidal syndrome' of dementia of the Alzheimer type.
Arch Neurol 1990; 47: 1318-1323

271. Van den Burg W, Van Zomeren AH, Minderhoud JM, et al.
Cognitive impairment in patients with multiple sclerosis and mild
physical disability.
Arch Neurol 1987; 44: 494-501

272. Van Gorp WG, Miller EN, Satz P, Visscher B.
Neuropsychological performance in HIV-1 immunocompromised
patients: a preliminary report.
J Clin Exp Neuropsychol 1989; 11(5): 763-73

273. Vanneste JAL, Hijman R.
Klinische en neuropsychologische kenmerken van 'normal pressure
hydrocephalus'. (Clinical and neuropsychological characteristics of
'normal pressure hydrocephalus.)
Ned Tijdschr Geneeskd 1987; 131: 1080-84

274. Von Sattel JP, Ferrante RJ, Stevens TJ, Richardson EP.
Neuropathologic classification of Huntington's disease.
J Neuropathol Exp Neurol 1985; 44: 559-77

275. Wallesch CW, Fehrenbach RA.
On the neurolinguistic nature of language abnormalities in
Huntington's disease.
J Neurol Neurosurg Psychiatr 1988; 51: 367-73

276. Warbuton DM.
Drugs and the processing of information.
In: Stahl SM, Iversen SD, Goodman EC (eds). Cognitive
neurochemistry.
Oxford, Oxford University Press, 1987: 111-34

277. Weinberger DR, Berman KF, Iadarola M, et al.
Prefrontal cortical blood flow and cognitive function in Huntington's
disease.
J Neurol Neurosurg Psychiatr 1988; 51: 94-104

278. Weintraub S, Mesulam M-M. Mental state assessment of young and
elderly adults in behavioral neurology.
In: Mesulam M-M (Ed). Principles of behavioral neurology.
Philadelphia, FA Davis Company, 1985: 71-123

279. Whitehouse PJ.
The concept of subcortical and cortical dementia: another look.
Ann Neurol 1986; 19: 1-16

280. Whitehouse PJ, Price DL, Struble RG, et al.
Alzheimer's disease and senile dementia: Loss of neurons in the basal
forebrain.
Science 1982; 215: 1237-1239

281. Wickelso C, Andersson H, Blomstrand C, Lindqvist G.
The clinical effect of lumbar puncture in normal pressure
hydrocephalus.
J Neurol Neurosurg Psychiatr 1982; 45: 64-69

282. Wilcock GK, Esiri MM, Bowen DM, Hughes AO.
The differential involvement of subcortical nuclei in senile dementia of
Alzheimer's type
J Neurol Neurosurg Psychiatr 1988; 51: 842-9

283. Wilson RS, Como PG, Garron DC, et al.
Memory failure in Huntington's disease.
J Clin Exp Neuropsychol 1987; 9: 147-54

284. Wisniewski HJM.
Progressive nuclear palsy.
In: Frederiks JAM (Ed). Handbook of Clinical Neurology, vol 2.
Amsterdam, Elsevier Science Publ BV, 1985: 301-03

285. Wolfson LI, Leenders KL, Brown LL, Jones T.
Alterations of regional cerebral blood flow and oxygen metabolism in
Parkinson's disease.
Neurol 1985; 35: 1399-1405

286. Yarchoan R, Thomas RV, Grafman J, et al.
Long-term administration of 3'-azido-2',3'-dideoxythymidine to
patients with AIDS-related neurological disease.
Ann Neurol 1988; 23 (suppl): S82-87

287. Yarchoan R, Berg G, Brouwers P, et al.
Response of human-immunodeficiency-virus-associated neurological
disease to 3'-Azido-3'-deoxythimidine.
Lancet 1987; 1: 132-35

288. Young AB, Penney JB, Starosta-Rubinstein S, et al.
PET scan investigations of Huntington's disease: cerebral metabolic
correlates of neurological features and functional decline.
Ann Neurol 1986; 20: 296-303

CHAPTER 4

Illustrative case histories

Introduction

The introduction of the concept of 'cortical' and 'subcortical' dementia has engendered considerable interest and debate in the past 17 years.

It has become clear that there exist not only cortical and subcortical dementia syndromes, but also combinations of the two [5]. Of clinical importance is the fact that many treatable dementias exhibit the syndrome of subcortical dementia [5,9]. This is not always true, as in the case of patients with acquired immunodeficiency syndrome (AIDS); a cortical dementia syndrome is most often caused by treatable cerebral opportunistic cerebral infections. The Aids Dementia Complex (ADC) is a subcortical type of dementia [7], and although antiviral treatment seems to halt progression [20], until now the clinical picture cannot be reversed. Dementia in central nervous system infections often shows mixtures of cortical and subcortical features. If diagnosed in time and given adequate treatment sometimes length of survival can be increased or even partial recovery can be reached [5].

The results of neuropsychological examinations do reveal differences between cortical and subcortical degenerative processes [8], and between localized cortical and subcortical lesions [11].

Three cases will be presented, illustrating the use of neuropsychological assessment in the examination of patients with signs and symptoms of a dementing disease.

Case histories

Case 1.
A 77-year-old grower of strawberries was admitted because of progressive gait disturbance, bradykinesia, impairment of balance, and urinary incontinence.

History
Two and a half years before admission he suddenly experienced a severe headache during the night. The next morning he seemed obtunded and had problems in walking. In the course of time these symptoms disappeared, although walking remained difficult. The headache attacks recurred and the clinical picture gradually became worse with progressive gait disturbances, bradykinesia, impaired balance, and urinary incontinence, and periodic drowsiness.

Neurological examination one month after the first headache attack revealed no abnormalities and a diagnosis of "basilar insufficiency" was made. Computer tomographic scanning, five months later, revealed generalized atrophy, without further abnormalities.

About one and a half year later a diagnosis of Parkinson's disease was made and bromocriptine was given. Six months later the patient was admitted in a hospital because the medication did not seem effective. On neurological examination he was inert and apathetic. No tremors were noted and cogwheel rigidity was absent. He walked with small steps and masked facies was noted. He was given L-Dopa in increasing doses, combined with bromocriptine. No significant changes were noted and the patient preferred to return home after about three weeks.

From then on, his condition deteriorated and he was put on the waiting list for a nursing home. One month after his discharge from the hospital, he was referred by his general practitioner to our hospital.

Neurological examination
On examination the patient was slow and disoriented. His blood pressure was 130/80. A minor postural tremor of the fingers was

present, a slight hypertonia of the arms and legs, and bilateral positive palmomental reflexes. His gait was broadbased with small shuffling steps. He was wearing a condom catheter for his incontinence. The tentative diagnosis after first examination was normal pressure hydrocephalus.

Ancillary investigations
Computed tomographic scanning without contrast showed considerable enlargement of both lateral ventricles and the third ventricle; after intravenous contrast a small lesion became apparent in the suprasellar region, probably a small meningioma. The electro-encephalogram showed diffuse slowing with intermittent bilateral frontal delta activity. Contrast injected in the cisterna magna filled the fourth and third ventricles, but was visible to a much lesser extent in the lateral ventricles only after 36 hours. The cerebrospinal fluid contained no pathological cells, with a protein concentration of 1.47 g/l.

Neuropsychological assessment
Neuropsychological assessment was performed four days after admission. The specific question to be answered concerned the presence and nature of the mental changes, e.i. 'cortical' and/or 'subcortical'. The test battery used was the same as in our investigation of mental changes in patients with subcortical arteriosclerotic encephalopathy [10], with a few additional tests.

The patient had only 6 years of formal education. His estimated premorbid level of intelligence was average. He was right-handed. He yawned frequently, and at the same time closed his eyes - which he slowly opened after a few minutes. He took a very long time to respond to questions or requests - even up to 8 minutes on for instance simple arithmetic items. After a short conversation it soon became clear that questions and instructions had to be very short - when too long the patient was not able to process the information adequately. This condition severely restricted the possibilities for a formal examination. Some insight in his cognitive difficulties was present; he admitted that he had become very slow, that thinking was difficult, and that his memory was not intact.

The present level of intelligence could not be reliably assessed. Based on his performance on several tests a deterioration seemed present.

This could be inferred from his score of 7 on both the Cognitive Capacity Screening Examination (CCSE: cut-off score 20/30)[12] and the Short Portable Mental Status Questionnaire (SPMSQ: cut-off score 27/30)[19]. His performance on a more extensive test, the Amsterdam Dementia Screening (ADS)[14], indicated mental deterioration. He scored below the cut-off values on all subparts of this test [15]: orientation in time, orientation in place, visual memory, auditory verbal memory, word fluency, copying drawings, and the Meander figure; in this last test perseverations occurred. Although right-left orientation was intact, his performance on a test with sequential items was hampered by perseverative behaviour. Memory for recent events, and to a lesser extent for remote events, was impaired. Auditory verbal immediate memory span was limited to four digits forward, while backward repetition was failed. Auditory verbal immediate and delayed recall were impaired; delayed recognition was relatively intact. Immediate recognition on a visual memory task (multiple choice recognition of five drawings) was intact; performance declined on repeated trials.

Initiation and rate of speech were very slow. Comprehension of spoken language (short sentences) and naming were intact. Oral use of language was limited to short sentences; syntaxis and word finding ability were normal. Category-bound word fluency (one minute) was limited to six animals and two professions.

The patient's writing was barely readable because of severe micrographia, with evidence of perseverations. Reading aloud was intact, although very slow. Because of the extreme slowness in responding, reliable assessment of reading comprehension was limited to very short sentences (maximum of four words). Basic arithmetic ability was intact. A short examination of praxis demonstrated no impairment.

Basic visual gnosis, as in recognition of objects and drawings, was intact. Copying of drawings was hampered by visuospatial impairment.

The profile of cognitive impairments pointed to *a subcortical type of dementia*. Performance was extremely slow. Mild perseverative behaviour was present. Auditory verbal memory recall was impaired, but recognition was relatively intact. Visuospatial ability was impaired. Basic reading and arithmetic ability were intact. There was severe micrographia. Aphasia and apraxia were absent.

Diagnosis

Based on the neurological examination, and ancillary investigations a diagnosis was made of obstructive hydrocephalus, caused by a small suprasellar meningioma. The subcortical dementia fitted in with this diagnosis, since this type of dementia has been described in patients with hydrocephalus [6,22].

Course

Ventriculocardial shunt surgery was performed. On neurological examination, three weeks later, a considerable improvement was noted. The patient was more alert and could walk better, and had no complaints of headache any more.

Considerable improvement was also demonstrated on repeated neuropsychological assessment. Speed of responding was faster. Scores on the CCSE and SPMSQ had improved to respectively 17 and 19. Performance on all subtests of the ADS had improved, although on three tests the scores were still below the cut-off values: auditory verbal memory, word fluency and Meander figure. Hardly any perseverations occurred on the right-left orientation task. On digit span he could repeat five digits forward, and three backward. His performance on the visual recognition memory task was normal. Auditory verbal memory performance, although improved, was still restricted on immediate and delayed recall. Category-bound word fluency had improved to respectively nine animals and seven professions in one minute. Handwriting, although still slightly irregular, approached near normal format. Initiation of speech and speech rate had improved. Performance on arithmetic and reading tasks was faster and better. Visuospatial ability was nearly normal.

A short neuropsychological evaluation, six months later,

demonstrated further improvement to near normal performance. The patient could hardly remember anything from the period preceding ventriculocardial shunting. He indicated that during that time, everything in his environment went too fast for him, and because of his slowness of thinking he could not keep up with it.

Case 2.

A 58-year-old university professor of mathematics was admitted to our hospital with a previously diagnosed progressive supranuclear palsy (PSP). The symptoms progressed rapidly and consisted of deteriorating balance and vision, hypersecretion of saliva, mental deterioration (mainly word finding difficulties), and bradykinesia.

History

The symptoms started about three and a half years earlier with intermittent dizziness. He fell off his bike two times, with short-lasting unconsciousness.

Two years later he was admitted to a hospital because of dizziness, diplopia and a tendency to doze off. The diagnosis myasthenia gravis was made.

One month later he was admitted to another hospital. At that time he experienced memory problems. A reduction in vertical gaze movements was noted. The electroencephalogram showed bilateral theta activity in the temporal and frontal regions, and delta activity more pronounced over the left than over the right hemisphere. Computed tomographic scanning revealed cerebral atrophy. On neuropsychological examination signs of organic deterioration were found, which was reported to be indicative for a dementing disease, e.g. Alzheimer's disease, and cortical atrophy. A diagnosis of probable Alzheimer's disease was made.

Two months later the patient was again admitted to another hospital. Based on the presence of mild to moderate signs of extrapyramidal dysfunction, vertical gaze paralysis and slowness noted on mental status examination, a diagnosis of progressive supranuclear palsy was made. He was given amantadine, which had no significant effects on his motor problems.

In the following five months his condition got worse. The

main symptoms were: deteriorating balance, stiff posture of the neck, progressive dizziness, a tendency to fall backwards, and fine motor impairment.

Six months after his last hospital admission, reading and writing became virtually impossible, and he had to quit his job. Again 6 months later he was admitted to a hospital. The diagnosis PSP remained unchanged. He was started on lisuride and trhihexyfenidylchloride and a mild improvement was noted in his gait and dexterity.

About one month later the patient visited our outpatient department. In the course of his disease he had been treated with several medications, such as L-dopa, amantadine and lysuride, all of which had little or no effects. About five months later he was admitted to our neurology department, because of further deterioration of his condition.

Neurological examination
The patient was alert and oriented. Spontaneous speech was slow and monotonous, with slight dysarthria. Examination of the cranial nerves demonstrated: bilateral lenticular cataract with reduced vision (1/2); a slight anisocoria (r<l); limited eye blinking movements; normal pupillary reactions to light; a vertical gaze paresis and saccadic horizontal pursuit; reduced optokinetic nystagmus in the horizontal plane, and reduced convergence reaction; brisk palmomental reflexes, and masseter reflexes; also a left-sided hypaesthesia in the maxillary distribution of the trigeminal nerve and a left-sided peripheral facial paresis, with blepharoraphy (after parotid surgery).

Motor examination showed: remarkable bradykinesia and hypokinesia, and masked facies; rigid posture; head held in flexion; a slight bilateral hypertonia without (cogwheel) rigidity or tremor; symmetric normal strength in legs and arms, and normal sensibility and coordination; symmetric reflexes and a bilateral extensor plantar response; rigid carriage when walking, with lack of arm movement and head in flexion; turning around performed 'en block'; and a negative Romberg test. While standing up he inclined to fall backwards.

Ancillary investigations
Computed tomographic scanning showed cortical atrophy and
minimal cerebellar atrophy. The laboratory evaluation and the
electrocardiogram were essentially unremarkable. The
electroencephalogram showed a generally lowered background
pattern, especially bilaterally in the temporobasal regions.

Neuropsychological examination
The estimated premorbid level of intelligence of this patient was
very high. He was righthanded.

About one and a half years before admission a neuropsychological assessment had been performed in another hospital,
with the following results: above average level of intelligence, as
measured by the Wechsler Adult Intelligence Scale (WAIS) [24],
which fell short of the expected (superior) level; spatial
impairment; defective concentration and memory; cognitive and
(psycho-)motor slowing; occurrence of perseverations; minor
language impairment, mainly consisting of dysarthria; impaired
judgment and critical attitude; and slight loss of decorum in his
behaviour. These observations and test results were interpreted as
indicative for an organic cerebral disorder fitting in with a
diagnosis of Alzheimer's disease and cortical atrophy.

Part of this examination was repeated in our hospital,
supplemented with additional language and memory tests, also
used in our investigation of the neuropsychological differentiation
between cortical and subcortical dementia [8]. The question was,
whether his cognitive status was in accordance with a diagnosis of
progressive supranuclear palsy.

During the examination the patient moaned frequently, and
although conscious of it, he could not stop it. Wordfinding
difficulties were apparent in conversational speech and speech was
dysarthric and monotonous, of which he was clearly aware. The
patient impressed as being very slow. His main complaints
concerned difficulties in talking, writing and reading, and to a
lesser extent slight difficulties in memorizing new information.

His level of intelligence (WAIS) was average, indicating
further deterioration. Speed of responding and performing
various tasks was slow. Concentration was reduced. Immediate

and delayed auditory verbal recall were impaired, both after single as well as repeated presentation, but delayed recognition was relatively intact. Analysis of his performance on these tasks indicated reduced ability to structure information. Performance on a visual memory task was severely restricted on immediate recall, delayed recall and recognition, with many perseverative responses, and this could not be accounted for by the patient's impaired vision.

In spoken language there were wordfinding difficulties, with semantic paraphasias, perseverations, intrusion errors, and also incidental neologisms. Syntaxis was defective. The patient could correctly name common objects and coloured pictures of objects, but performance deteriorated on more complex photographs and (black-and-white) line drawings. Category-bound word fluency was severely restricted, and occasionaly intrusion errors were present on these tasks. His verbal abstract reasoning ability was impaired, and his performance indicated difficulties in organizing thought activities. Despite these language disturbances, he made no errors on the Token test, although he took a very long time to respond.

Basic visual gnosis was intact. Visuospatial ability was slightly impaired, which could partly explain the defective performance on visuoconstructional tasks. Performance on these tasks was also hampered by defective structuring ability.

Basic arithmetical ability, examined with simple, discrete problems was intact. However to his own distress this mathematician made many mistakes on more complex arithmetical tasks; in trying to correct them he only made further mistakes. Part of this impairment could be explained by concentration and memory difficulties. A severe micrographia made standard examination of writing impossible - no errors were present in the few readable words. The patient could read and understand text in large print. Performance on tasks measuring ideational and ideomotor praxis was normal, except for minor coordination difficulties caused by his restricted vision.

Tactile recognition of common objects with the left hand was impaired; in addition he made significantly more mistakes in naming objects placed in the left hand compared to the right hand.

Recognition of sandpaper figures [3] was defective for both hands.

During examination the patient showed remarkable cognitive and motor slowing, and a reduced ability to organize thought and motor activities.

It was concluded that there were clear indications for the presence of *a subcortical dementia, accompanied by cortical neuropsychological deficits.*

The indications for subcortical dementia were: memory impairment with relatively normal recognition; cognitive and motor slowing; occurrence of perseverations; impaired ability to manipulate acquired knowledge; dysarthria and monotonous speech; reduced word fluency; occurrence of intrusion errors on language tasks; micrographia; visuospatial impairment; absence of apraxia and alexia. These deficits partly resembled the characteristics of the dementia in patients with progressive supranuclear palsy (PSP) [1,6].

The cortical neuropsychological deficits consisted of: paraphasias in complex naming tasks and in spoken language (with defective syntaxis); tactile agnosia; the severity of perseverative behaviour and of the impaired abstract reasoning and judgment ability were also thought to be indicative of a cortical mainly frontal lobe disorder. These signs of cortical impairment has not been reported in patients with PSP.

Diagnosis
Based on the neurological examination, the diagnosis of progressive supranuclear palsy remained unchanged. Although the cognitive profile of a subcortical type of dementia fitted in with this diagnosis, the presence of cortical neuropsychological deficits did not.

Course
After his discharge from the hospital the patient's condition gradually worsened. He became completely incapable of functioning independently, and needed 24 hour skilled nursing care. He also became frequently incontinent for both urine and faeces, and his periods of coherence became less and less. He died about four and a half years after the onset of the symptoms.

Post-mortem examination of the brain demonstrated frontal atrophy. In addition, the pons and mesencephalon impressed as being small, and the ventricles were slightly enlarged. There were minor arteriosclerotic changes of the large cerebral arteries. Microscopic examination revealed: a minor loss of neurons in the frontoparietal area; loss of neurons in the substantia nigra (with reactive gliosis); perivascular lymphocytic infiltrates in the brain stem and around the third ventricle; focally, some microglial nodules were found and some brainstem neurons showed intranuclear eosinophilic inclusion bodies (type A Cowdry). There were no indications for abnormalities compatible with the diagnosis of progressive supranuclear palsy. It was concluded that the demonstrated abnormalities were compatible with a diagnosis of a chronic viral brainstem infection of unknown origin, and minor (frontoparietal) cortical atrophy.

The results of the neuropsychological examination were in accordance with the neuropathological findings of subcortical and (predominantly fronto-parietal) cortical abnormalities.

Case 3.

A 29-year-old homosexual man was admitted to our hospital, because of symptoms of slowing in rate of speech, thinking and movement, and impairment of writing.

History

His first symptoms dated back 15 months. The patient then experienced a floating feeling in his head, without vertigo or any other complaints.

Four months later his speech became quavering and slower. He was seen by a neurologist, who noted no neurological abnormalities. Computed tomographic scanning was unremarkable. Again four months later he twice experienced a sudden and short-lasting flapping of both upper arms, without other signs. Because of the gradual progression of his complaints he was admitted to a hospital one month later. Extensive neurological examination demonstrated no abnormalities.

Six times during the next month the patient experienced short-lasting (10 seconds) attacks of being unable to speak. During these attacks he seemed absent-minded. Gradually his ability to concentrate seemed to decline, and his speech and handwriting deteriorated.

One year after the onset of the symptoms he was again referred to a neurologist. Computed tomographic scanning showed cerebral and cerebellar atrophy. Examination of the cerebrospinal fluid indicated increased protein and IgG-index. After this the patient was referred to our hospital.

His symptoms now consisted of impaired and slow speech, impaired handwriting, mental and physical motor slowing, and memory problems.

Neurological examination

General physical examination showed no abnormalities. The patient seemed slightly bradyphrenic, and his voice was mildly tremulous. There were no signs of meningeal irritation.

Cranial nerve examination demonstrated bilateral pale papillae; no further neuroophthalmic and oculomotor abnormalities; an irregular twitching movement in the face with a small amplitude; no paresis of the facial nerve; tremulous protruding of the tongue, but normal tongue movements; and a brisk masseter reflex.

In both arms a fine irregular rest- and intention tremor was present. There was no arm drift. Motor strength and sensibility were unremarkable. A bilateral dysdiadochokinesis was present (right>left). Reflexes were symmetrical and brisk.

Muscular tone, strength, and sensibility of the legs were unremarkable. The heel-to-shin-test was impaired. Bilaterally, reflexes were very brisk and symmetrical, with extensor plantar reflexes. Gait was slightly irregular, but turning around caused no difficulties. He could walk in a straight line (heal-toe-walking), and Romberg's test was normal.

Ancillary investigations

Serum VDRL was 1:4; TPHA >1:40960. Tests for antibodies to HIV were negative. The cerebrospinal fluid showed 147 mononuclear and 2 polymorphonuclear cells per 3 mm^3. Total protein and IgG and IgM were increased. VDRL was negative;

TPHA 1: 20480, and the FTA-absorption test was positive.

Chest X-ray and electrocardiogram were normal. Cerebral MRI scanning demonstrated central and peripheral atrophy with minor periventricular lining, without appreciable white matter abnormalities.

The ophthalmologist noted bilateral temporal pale papillae, and otherwise no abnormalities, especially no vasculitis, chorioretinitis, or Argyll-Robertson pupils. Visual field examination was essentially unremarkable. Left and right visual evoked responses were normal.

Diagnosis
The patient was diagnosed as suffering from neurosyphilis, with dementia (general paresis).

Neuropsychological examination
Although the patient had a negative HIV-test 7 months before admission, the possibility of mental changes as described in patients with ADC [7] was not completely excluded. The main purpose of the neuropsychological examination was to provide a starting point for therapy evaluation.

The patient had 13 years of formal education. He was a clerical worker on the film department of a broadcasting company. He was righthanded, and colour-blind (red-green). His favourite pastime consisted of acting in an amateur theatre group; because of his symptoms he had given up on this in the year preceding his admission. His main complaints concerned his speech, trembling of his hand, a slowing in performing activities, and loss of facial expression. Apart from some minor concentration problems, he denied the presence of any other cognitive difficulties.

During the examination the patient showed very little insight in his cognitive impairment. He made a childish impression, and his affect was flat. A slight tremor of his voice and both hands was noted.

The neuropsychological test-battery was the same as used in an evaluation of mental changes in patients with AIDS [7], supplemented with a few tests.

His scores on the Mini Mental State Examination and

Cognitive Capacity Screening tests were 25 and 24 respectively; errors were made on items with regard to attention and memory. The orientation in time, in place, in person, and in space was unremarkable. On a task for right-left orientation he made perseverative errors.

The level of intelligence had deteriorated from average (estimated premorbid level) to far below average, primarily caused by language related deficits, poor arithmetical performance, and an inability to carry out a visual sorting subtest.

Psychomotor speed was very low, and deficits of attention and concentration were present.

Immediate auditory verbal memory span was limited to five digits forward and three digits backward. Immediate and delayed auditory verbal recall were impaired, but delayed auditory verbal recognition was relatively unaffected. Visual immediate and delayed reproduction of figures, and visual recognition of faces (Recognition Memory Test) [23] were impaired.

Word fluency was restricted, due mainly to slow responding but also no improvement was found when the time limit was extended. Features of aphasia were absent, although knowledge of words was impaired. Apart from the tremor in his voice, reading aloud was unremarkable, as was reading for comprehension.

A tremor was visible in the patient's handwriting, further characterized by irregular size of individual letters. Although his basic arithmetic ability was intact, many errors occurred on oral and written tasks, primarily caused by defective planning and limited concentration. No improvement in performance was noted when time limits on tasks were extended.

A short examination of ideomotor and ideational apraxia and tactile gnosis was normal. Performance on a line bisection test and a line orientation test [3] indicated a minor right-sided inattention and visuospatial impairment. Visual gnosis of objects, photographed from normal and unusual angles, was normal. His visual organization ability was moderate.

A tremor was visible in graphomotor performances and handwriting was irregular. His drawings tended to be small, and details were copied in an abnormal order. The patient was unable to draw a cube; placement of the numbers and the hands in a clock indicated visuospatial impairment. Performance on block construction (WAIS-R) was severely defective; the way he tackled the items indicated both a visuospatial deficit and especially defective organization and planning. No improvement was noted when these tasks were subdivided and offered in parts.

Reasoning ability, concept formation and planning of activities were (severely) defective; also on some tasks perseverative errors occurred. Again no improvement was noted when structured help was given to perform these tasks.

Scores on mood questionnaires [17,21] did not indicate changes; he only felt less vigorous than before.

It was concluded that this patient's profile of cognitive function pointed to *a combination of cortical and subcortical dementia*. The cortical neuropsychological deficits mainly consisted of: severe deficits on tasks requiring sorting, concept formation and planning, with frequent occurrence of perseverations and no improvement in performance when time limits were extended of structured help was given; impairment on language related tasks; no improvement in performance on word fluency tasks when the time limit was extended [8]. This was accompanied by limited insight. This combination primarily indicated to frontal cortical dysfunction. The presence of subcortical neuropsychological deficits was concluded from impairment of: attention and concentration; visual and auditory verbal memory, with relatively intact auditory verbal delayed recognition after repeated presentations; visuospatial ability; and psychomotor speed. This was accompanied by graphomotor deficits and reduced facial expression [8].

The cognitive profile did not fit in with a diagnosis of Aids Dementia Complex (ADC). In patients with ADC the prominent features are deterioration of: level of intelligence; memory; attention and concentration; psychomotor speed; memory; and (in many cases) wordfluency. Only mild impairment of executive control functions have been found. Visual information processing ability may be slightly defective, but this is usually a sequela of abnormalities in the visual system. Features of cortical dysfunction are absent [6,7].

The neuropsychological findings were in accordance with a

diagnosis of neurosyphilis with dementia. This central nervous system infection produces a mixture of cortical and subcortical features [5]. The cortical features most frequently suggest frontal lobe involvement [16,18].

Course
The patient was treated with intravenous penicillin during 14 days. Clinical improvement was already noted during this period.

Three months after treatment the neuropsychological examination was repeated. A slight improvement was noted with regard to: level of intelligence; memory; visuo-motor speed; visuospatial ability; abstract reasoning and judgement; word fluency; writing and arithmetic. No perseverations were present any more. The tremor in handwriting and drawing had almost completely disappeared. A slight residual dysarthria was noted. Affect and facies were unremarkable. Mood was normal. His insight in his - not yet completely recovered - cognitive functioning was still slightly limited.

Discussion

Three cases are presented to demonstrate the use of neuropsychological assessment in the examination of patients with dementing diseases. The first patient was thought to have an (obstructive) hydrocephalus. Normal pressure hydrocephalus is a potentially reversible condition, accompanied by the triad of fluctuating but progressive dementia, (urinary) incontinence and gait disturbance [2,13]. The dementia is of a subcortical type [5,6,22]. Alzheimer's disease may also be associated with hydrocephalus, caused by atrophy of the brain, producing compensatory enlargement of the ventricular system. Shunting in these patients will not lead to improvement [5]. The results of the neuropsychological examination of this patient were in accordance with a diagnosis of subcortical dementia and not with that of Alzheimer's disease. The patient underwent ventriculocardial shunting, and in time improved significantly to near normal performance.

The second patient was admitted to our hospital with a

diagnosis of progressive supranuclear palsy (PSP) and progressive symptoms. The dementia in PSP is one of the 'classic' examples of subcortical dementia [1]. A notable feature is the absence of cortical features such as aphasia, apraxia and agnosia. The results of the neuropsychological examination of this patient showed both subcortical and cortical features of dementia. This picture was thought to be incompatible with an exclusive diagnosis of PSP. Post-mortem examination, performed one year later, demonstrated the presence of subcortical abnormalities, most consistent with a diagnosis of a chronic viral brain stem infection, and minor (frontoparietal) cortical atrophy.

The third patient was admitted to our hospital with slowness of speech, mental and physical slowing, and impaired writing. Based on neurological examination and ancillary investigations a diagnosis of neurosyphilis with mild dementia was made (general paresis). Syphilitic dementia ('Dementia paralytica') is a potentially reversible disorder [5,16]. Neuropsychological examination was performed to provide a baseline for therapy evaluation, and for evaluation of the cognitive characteristics of his mental deterioration. The cognitive profile showed features of subcortical and mainly frontal cortical dysfunction, and was different from that of patients with Aids Dementia Complex (ADC) [7]. The presence of a mixture of cortical and subcortical features of dementia is present in patients with dementia paralytica, with often prominent signs of frontal cortical dysfunction [5,16,18]. Three months after treatment he showed improvement on all aspects of the neuropsychological examination.

These three cases illustrate the possibilities to differentiate between cortical and subcortical dementia, and dementing diseases with mixed features. This distinction is relevant, as some conditions are reversible or their progression can be halted [5,7,9].

Neuropsychological examination may assist in making the right diagnosis. In the past 17 years the number of publications with regard to the concepts of cortical and subcortical dementia has steadily increased. The focus is mainly on the prototypical examples, e.g. Alzheimer's disease, progressive supranuclear palsy, Parkinson's disease, and Huntington's disease. Until now, less attention has been given to diseases with mixed cortical and subcortical dementia, especially with regard to the cognitive features.

The patients described in this chapter were examined with specially constructed test batteries. The results of their performance demonstrate the possibility of neuropsychological differentiation of dementia syndromes, important in the diagnostic analysis of patients with dementia [4].

Acknowledgement

I wish to thank dr. JD. Speelman MD., dr. J. van Manen MD., dr. J. de Gans MD., dr. J. Stam MD., and J. Tielens MD., who performed the neurological examinations, and dr. D. Troost MD., neuropathologist who performed the postmortem examination of the second patient. My special thanks go to dr. A Hijdra MD., who helped to summarize the results of the neurological examinations and ancillary investigations of the three patients.

Refereences

1. Albert ML, Feldman RG, Willis AL.
    The 'subcortical dementia' of progressive supranuclear palsy.
    J Neurol Neurosurg Psychiat 1974; 34: 121-30

2. Benson DF.
    Hydrocephalic dementia.
    In: Frederiks JAM (Ed). Handbook of clinical neurology, vol 2.
    Amsterdam, Elsevier Science Publ BV, 1985: 323-33

3. Benton AL, Hamsher K de S, Varney N, Spreen O.
    Contributions to neuropsychological assessment.
    New York, Oxford University Press, 1983

4. Crevel H van.
    Clinical approach to dementia.
    In: Swaab DF, Fliers E, Mirmiran M, et al (Eds). Aging of the brain and senile dementia; Progress in Brain Research, 70.
    Amsterdam, Elsevier Science Publ BV, 1986; 70: 3-13

5. Cummings JC, Benson DF.
    Dementia: a clinical approach, sec. ed.
    Boston, Butterworth, 1992

6. Derix MMA.
    Dementia in diseases predominantly affecting subcortical structures: a
    neuropsychological contribution to the concept of subcortical dementia.
    This book, chapter 3

7. Derix MMA, de Gans J, Stam J, Portegies P.
    Mental changes in patients with AIDS.
    Clin Neurol Neurosurg 1990; 92: 215-22

8. Derix MMA, Groet E.
    Corticale en subcorticale dementie: een neuropsychologisch onderscheid.
    (Cortical and subcortical dementia: a neuropsychological distinction.)
    In: Schroots JJE, Bouma A, Braam GPA, et al (Eds). Gezond zijn is
    ouder worden.
    Assen, van Gorcum, 1989: 179-88

9. Derix MMA, Hijdra A.
    Corticale en subcorticale dementie, een zinvol onderscheid? (Cortical
    and subcortical dementia, a useful distinction?)
    Ned Tijdschr Geneeskd 1987; 131: 1070-3

10. Derix MMA, Hijdra A, Verbeeten BWJ.
    Mental changes in subcortical arteriosclerotic encephalopathy.
    Clin Neurol Neurosurg 1987; 89: 71-8

11. Fromm D, Holland AL, Swindell CS, Reinmuth OM.
    Various consequences of subcortical stroke.
    Arch Neurol 1985; 42: 943-50

12. Jacobs JW, Bernhard MR, Delgado A, Strain JJ.
    Screening for organic mental syndromes in the medically ill.
    Ann Int Med 1977; 86: 40-46

13. Katzman R.
    Normal pressure hydrocephalus
    In: Wells CE (Ed). Dementia. 2nd ed.
    Philadelphia, FA Davis, 1977: 69-92

14. Lindeboom J, Jonker C.
    ADS-6: Amsterdamse Dementiescreening-6. Handleiding. (Amsterdam
    Dementia-screening-6. Manual.)
    Lisse, Swets & Zeitlinger, 1989

15. Lindeboom J, Jonker C.
    Neuropsychologisch onderzoek van bejaarden: een follow-up studie.
    (Neuropsychological assessment of the elderly: a follow-up study.)
    T Psychiatr 1980; 22: 498-503

16. Lishman WA.
    Organic psychiatry. The psychological consequences of cerebral
    disorder. 2nd ed.
    Oxford, Blackwell Scientific Publications, 1987

17. McNair DM, Lorr M, Droppleman LF.
    E.I.T.S. Manual for the Profile of Mood States.
    San Diego, Educ. and Industr. Testing Service, 1971
    Dutch revision: Wald F. De verkorte POMS. Thesis. Amsterdam,
    University of Amsterdam, 1984

18. Michel D, Blanc A, Laurent B, et al.
    Etude biologique, psychométrique et tomodensitométrique de la
    neurosyphilis traitée.
    Rev Neurol (Paris) 1983; 139: 737-44

19. Pfeiffer E.A.
    A short portable mental status questionnaire for the assessment of
    organic brain deficits in elderly patients.
    J Am Geriatr Soc 1975; 23: 433-41

20. Portegies P, de Gans J, Lange JMA et al.
    Declining incidence of AIDS Dementia Complex following introduction
    of zidovudine treatment.
    Br Med J 1989: 299: 819-21

21. Spielberger CD, Gorsuch RL, Lushene RE.
    STAI Manual for the State-Trait Anxiety Inventory.
    Palo Alto, Consulting Psychologist Press, 1970
    Dutch revision: Ploeg van der HM, Defares PB, Spielberger CD. Zelf-
    Beoordelings Vragenlijst (ZBV): een Nederlandstalige bewerking van de
    Spielberger State-Trait Anxiety Inventory.
    Lisse, Swets & Zeitlinger BV, 1980

22. Vanneste JAL, Hijman R.
Klinische en neuropsychologische kenmerken van 'normal pressure hydrocephalus'.(Clinical and neuropsychological characteristics of 'normal' pressure hydrocephalus.)
Ned Tijdschr Geneesk 1987; 131: 1080-84

23. Warrington EK.
Recognition memory test: manual.
Windsor, NFR-Nelson, 1984

24. Wechsler D, Stone C.
Manual for the Wechsler Adult Intelligence Scale.
New York, The Psychological Corporation, 1955
Dutch revision: Stinissen J, Willems PJ, Coetsier P, Hulsman WLL.
Handleiding bij de Nederlandstalige bewerking van de WAIS.
Lisse, Swets & Zeitlinger, 1970

CHAPTER 5

Neuropsychological assessment in dementia
Guidelines

Introduction

In recent years the objective of clinical neuropsychological assessment has shifted from the prediction of the presence or site(s) of brain damage to the assessment of behavioural change and the delineation of impaired and unimpaired cognitive functions [3,19,54,57]. The focus of human neuropsychology has shifted from 'topography' to a concern with underlying neural and cognitive mechanisms [3,81]. The benefits of a thorough neuropsychological evaluation in the examination of patients with (possible) dementia include [27,28,55,71]: description of the extent and quality of cognitive and possible emotional dysfunction; assistance in the formulation of a differential diagnosis and the prediction of the course of the disease; provision of a baseline level of cognitive functioning for therapy evaluation; and provision of the patient's remaining potential to advise caretakers and members of the patient's family.

It has been shown that the neuropsychological profiles of various dementia syndromes are quite distinct. Not only can a distinction be made between cortical dementia, subcortical dementia and mixed forms [22,23,25,26,28,29], but also there is evidence that (subtle) differences exist between the dementia syndromes within these categories [9,14,33,41,53,62-64,67,68,72,78].

Cortical and subcortical dementia: general characteristics

Cortical dementias generally produce deficits in learning and memory, language-abilities, visuoperceptual and visuoconstructional skills, and other intellectual functions such as arithmetic 14,26,71. This results in clinical entities such as amnesia, aphasia, apraxia, agnosia and acalculia. Alzheimer's disease and Pick's disease, sharing many features, are the two principal cortical dementias 14. But there are differences in clinical phenomenology (including neuropsychological findings), and in general the diseases are most readily separated on the basis of early behavioural changes and the occurrence of deficits in the temporal course of the disease 14, although a recent publication reports that the cognitive profile in patients with Pick's disease is variable and sometimes difficult to distinguish from that of patients with Alzheimer's disease 53. Changes of mood, and strange behaviour with sometimes the occurrence of the Klüver-Bucy syndrome are among the earliest signs of Pick's disease, while in early Alzheimer's disease social skills and personal care are relatively preserved 14,87. Amnesia, visuospatial disorientation, and acalculia are early features of Alzheimer's disease and generally occur only in later stages of Pick's disease 14,53. Language disturbances occur in both diseases; empty speech and reduced word fluency is characteristic for early Alzheimer's patients, and in early Pick's disease language impairment is relatively minor with semantic anomia and circumlocution more prominent 14,87; in later stages of the disease Alzheimer patients develop characteristics of a fluent aphasia, while the use of language of patients with Pick's disease is reminiscent of frontal lobe impairment 14,87. Alzheimer's disease can also be distinguished from so called 'frontal lobe dementia', which in some cases may represent forms of Pick's disease 67, although frontal lobe dementia is also found in dementia lacking distinctive histologic features, apart from cell loss and astrocytosis 52. Recently, different subtypes of dementia of the Alzheimer type have been distinguished based on the progression of the disease, the severity and pattern of intellectual and functional decline, accompanying motor disturbance and age of onset 64,68; also subgroups of Alzheimer patients with qualitative different neuropsychological profiles have been described [62]. A syndrome of progressive aphasia has been described in patients in whom general cognitive impairment became only manifest several years after onset of the language impairment. Postmortem examination was performed in two patients; characteristics of Alzheimer's disease were present in one, and the other displayed widespread neocortical changes without specific histopathological markers [37]. The syndrome of progressive aphasia has also been described in a patient with Pick's disease [36], a patient with Creutzfeldt-Jakob disease [61], and in patients with focal spongiform degeneration [51].

The syndrome of subcortical dementia is clinically characterized by bradyphrenia, forgetfulness, impaired cognition, apathy and often depression [14,18]. Impaired performance is found on neuropsychological tasks requiring: attention; effort; cognitive speed; processing, encoding and manipulating information; and executive control functions [22]. Features of aphasia, apraxia and agnosia are absent [22,26]. Within this general description however subtle differences can be found between various diseases with subcortical dementia. For instance in one study cognitive and behavioural impairment was compared between patients with progressive supranuclear palsy, patients with Parkinson's disease and Alzheimer patients with a comparable degree of intellectual deterioration; the scores of patients with Parkinson's disease on verbal and visuospatial tests were slightly better than those of patients with progressive supranuclear palsy. The latter group performed also slightly worse on frontal lobe tests and exhibited more frontal-type behaviour than patients with Parkinson's disease [72]. Matched groups of patients with multiple sclerosis and of patients with Huntington's disease could be distinguished on account of memory and arithmetical impairment, which were more severe in the latter group, although the overall performance on a range of neuropsychological tests was similar in both groups [9]. Compared with patients with Parkinson's disease, patients with progressive supranuclear palsy show more frontal lobe type impairment and are more impaired on tasks for clinical assessment of cognitive slowing [33]. A study examining motor learning and lexical priming, showed that demented Parkinson patients were impaired on both tasks, while demented Huntington patients were only impaired on the first task [41].

Neuropsychological assessment in dementia: issues raised

The majority of patients with a dementia syndrome are 65 years or older. A good deal of controversy exists as to exactly what cognitive changes occur as a function of the normal aging process, and conflicts over cross-sectional versus longitudinal designs hamper the interpretation and generalization of many studies [71]. A related problem concerns the definition and the assessment of the severity of dementia [22]. A large source of diagnostic error is caused by the difficulty in distinguishing between dementia and depression, especially in older age groups [5,42,49,83]. Neuropsychological assessment and interpretation of the data in the older age group is confronted with methodological problems and problems related to reliability of assessment [2,5,27,34,71]. The main methodological difficulties are: a lack of age-appropriate norms for various tests and the impossibility to discriminate low level scores; absence of adequate control groups, an unequal number of males and females, and insufficient provision for differences in level of education and in (premorbid) level of intelligence in many studies. Also the definition of many 'healthy elderly control groups' is often insufficient as many chronic diseases such as hypertension [84], diabetes, pulmonary disorders and cardiac diseases can have a negative influence on cognitive function. Problems related to reliability of assessment concern the comparison between test results of older and younger subjects such as: leaving aside differences in motivation; neglecting adverse effects of fatigue and the fact that elderly subjects are less accustomed to test situations [27].

An important problem in studying cognitive function in patients with dementia concerns the definition of dementia and the assessment of the severity of dementia. Patients with limited cognitive impairment most often do not meet the criteria for dementia incorporated in the definition of the Diagnostic and Statistical Manual of mental disorders (DSM-III-R)[32]. The operational definition of dementia by Cummings and Benson [14] does provide for inclusion of patients with less prominent intellectual deficits. Severity of dementia is often measured by the patient's performance on short mental state screening tasks, or is rated according to the presence of cognitive and behavioural deficits. Many of these instruments were mainly developed for distinguishing and/or evaluating patients with Alzheimer's disease, and concentrate on the presence of cortical deficits, and therefore are often unsuited for grading patients with subcortical dementia [22,24-26,28,29].

Signs of memory dysfunction and other changes in cognitive functioning are prominent not only in patients with dementia, but also in patients suffering from depression [27,42,49,71,83]. Especially in older patients the dementia syndrome of depression can occur, though it has been known to occur in younger patients as well. Standard neuropsychological tests can reflect cognitive impairment in patients with the dementia syndrome of depression. The most frequently reported impairment concerns: poor performance on motor tasks, deficits in attention and (cognitive) processing speed, memory deficits [46,49,50,71]. This profile of cognitive dysfunction is partially reminiscent of the type of impairment found in patients with subcortical dementia, who also often manifest personality changes such as irritability, apathy or depression. Signs and symptoms of depression can herald the onset of dementia or they can be accompanying features of a developing dementia [27,42,49]. Based on detailed history-taking, quantitative and especially qualitative neuropsychological evaluation of patients with depression and cognitive signs and symptoms, the following findings have been reported [27,46,49,50,71,83]: the onset of cognitive difficulties is often relatively short in terms of weeks or months, and the patient can often give an estimation of it; a depressive patient often complains more about his difficulties and has a better insight than patients with dementia; contrary to patients with dementia, there is often a incongruity between behaviour and apparent severity of the (reported) cognitive deficit in depressed patients; compared with demented patients, depressed patients will respond more often with "don't know" answers, give less false-positive and good-positive answers, and are less inclined to guess and to give near-miss answers; depressive patients will often show a disproportionate increase in errors on tasks requiring effort, and in general they will give up more readily on relatively simple tasks on which they will also make more mistakes than patients with dementia; performance of patients with depression on a range of tasks of comparable difficulty often shows great variability.

Despite the problems discussed, careful examination of clearly defined patients with dementia has revealed differences between cortical and subcortical dementia and mixed forms, and these findings can be used in the clinical neuropsychological assessment.

Neuropsychological assessment: cognitive functions and tests

The determination of a dementia syndrome should be made in the context of a multidisciplinary approach, using well-defined criteria in an assessment of multiple cognitive domains, the outcomes of which have to be interpreted in the light of available medical records. Because of physical or sensory handicaps, advanced age and sometimes reduced cooperation, often a flexible kind of procedure has to be used, in which appropriate tests are chosen according to the individual patient's condition [82]. In *table 1* a summary is given of the neuropsychological features of cortical and subcortical dementia.

Mental state screening
A careful evaluation of a patient with (possible) dementia must include: a detailed interview with the patient and if possible a family member, and a quantitative and qualitative assessment of cognitive (and emotional) functioning, general physical and neurological examination; attention has to be paid to state of awareness, general appearance and behaviour, and mood and affect [1,13,14,71]. Part of this evaluation can be performed with the use of extensive dementia screening tests such as the recently published Cambridge Examination for Mental Disorders of the Elderly (CAMDEX)[76,77], which also includes items related to the diagnosis of clouded/delirious state, depression and other psychiatric symptoms. The CAMDEX incorporates an extensive interview with the patient and with the caregiver, a cognitive

Table 1.   Neuropsychological features of cortical and subcortical dementia.*

	Cortical dementia	Subcortical dementia
Mental status screening	below cut-off	relatively good
Intelligence	deterioration, due to aphasia, amnesia, acalculia apraxia,visuo- perceptual, and visuospatial deficits	mild deterioration due to cognitive slowing, reduced attentional capacity, and defective executive control functions
Memory		
– remote memory	temporal gradient	flat gradient
– immediate recall	impaired	impaired
– delayed recall	impaired	impaired
– recognition	impaired	relatively normal
– learning capacity	impaired	relatively normal
– procedural learning	normal	impaired
Perceptual and psychomotor speed	normal; if impaired then due to other cognitive deficits	slowed down
Attentional activities	impaired, except for auditory attentional span	impaired, except for auditory attentional span **
Executive control functions	impaired; no/little benefical effects of help given	impaired; benificial effects of extra structuring or other help given
Language		
– speech	normal	impaired
– naming	often anomia	normal or (very) mild anomia
– word fluency	impaired	reduced - slow word finding
– comprehension	impaired	normal or mildly impaired
– repetition	normal	normal
– handwriting	normal	impaired
– linguistic aspect of writing	impaired	normal
– reading aloud	often impaired	normal, except for speech abnormalities
– reading comprehension	impaired	normal
Arithmetic	impaired, due to acalculia, visuo-spatial or aphasic deficits	normal or mildly impaired, due to planning deficit and/or use of time limits.
Praxis		
– ideomotor praxis	impaired	normal
– ideational praxis	(mildly) impaired	normal
Visual information processing		
– (basic) visual recognition	normal or mildly impaired	normal
– visual organization	impaired	normal or mild, task specific impairment
– visuospatial ability	often impaired	normal or mild, task specific impairment
Visuoconstruction	impaired, often due to visuo-spatial deficits	mildly impaired, often due to planning/structuring deficits
Motor Speed	normal	slowed down
Mood	relatively normal	depression, apathy, inertia, irritability

*   (Examples of) tests are described on page 158-166
** Auditory attentional span is often impaired in patients with progressive supranuclear palsy
      (PSP) and patients with Huntington's disease (HD)

examination (Cambridge Cognitive Examination - CAMCOG), an observational scale, and registration of findings of physical and neurological examination, and ancillary investigations. In the Dutch version of this test (CAMDEX-N) items have been added to this last part, partly in view of the importance of these findings for the differential diagnosis of various dementia syndromes [12,24]. Various cognitive dementia screening tests currently in use in the Netherlands and the Dutch-speaking part of Belgium, are described by Gilson et al. (1990)[35], and attention is paid to the conditions under which such a examination should take place.

To differentiate between early stages of various dementia syndromes, the use of cognitive screening tests is often insufficient and has to be supplemented by a more extensive neuropsychological evaluation in which at least special attention has to be paid to: intelligence, memory, perceptuo-motor speed, attention, executive control function, language and language related functions, arithmetic, praxis, visuoperception, visuoconstruction, motor speed and mood [22,23,25,26,29].

Intelligence
The general level of intelligence can be measured with tests such as the Wechsler Adult Intelligence Scale (-Revised), Raven Progressive Matrices and Groninger Intelligence Test [57,59,90]. A possible deterioration can only be inferred from a comparison with the estimated premorbid level [19,59]. This estimation is sometimes done on the basis of the patient's performance on so-called 'hold' tests, i.e. tests which are relatively resistant to the effects of ageing and early stages of diffuse cerebral disorders. Performance on the New Adult Reading Tests (NART)[69] correlates well with overall ability level and tends to resist the dementing process; a Dutch version has been published [79]. An other approach is the use of regression formulas in which account is taken of the subject's age, sex, level of education and occupation [91]. Pattern-analysis, error-analysis and comparison of performance on timed and non-timed tasks can give valuable information as to the reasons for intellectual decline.

Memory
Examination of memory function is of special importance in the
evaluation of dementia. One has to distinguish between remote
memory, immediate and delayed recall, and recognition. If
possible, learning capacity over repeated presentations of the
material to be remembered and procedural learning should be
examined. At least the auditory (verbal) and visual modality
should be examined [22,23,26,45,49]. Many different tests exist for
examining memory function, and descriptions can be found in
Lezak (1983)[57], Weintraub & Mesulam (1985)[90], Kapur (1988)[48],
Deelman (1990)[19]; examples include subtests of the well-know
Wechsler Memory Scale (revised) (WMS(-R)), Rey auditory verbal
learning and its Dutch equivalent, 'the 15-Word test'[19],
Recognition Memory Test [30,89]. Memory is impaired in patients
with subcortical as well as in patients with cortical dementia
[22,23,25,26,28,29,45]. Unlike patients with Alzheimer's disease however,
patients with subcortical dementia do show an increase in
immediate and in delayed recall of items on repeated
presentations, and also when the interstimulus interval of for
instance the 15-word test and logical memory subtest of the
Wechsler Memory Scale are prolonged [26]. Also verbal auditory and
visual delayed recognition, and the rate of forgetting are relatively
normal in patients with subcortical dementia [22], contrary to
patients with signs and symptoms of cortical dementia [48].
Performance on remote memory tasks shows a temporal gradient
in patients with Alzheimer's disease, while impairment of remote
memory is equally severe for all decades in patients with
Parkinson's disease, Huntington's disease and multiple sclerosis
[45]. Procedural learning is defective in patients with Parkinson's
disease and patients with Huntington's disease [22] contrary to
normal performance in patients with Alzheimer's disease [8].

Perceptual and psychomotor speed
In evaluating test results of individual patients one should pay
attention to the qualitative analysis of their performance. Many
neuropsychological tests are characterized by the use of time limits
and exceeding the time limit is different from not being able to
perform the task. Slowed information processing, e.g. perceptual

scanning speed, reaction time, and psychomotor speed, is a cardinal feature of subcortical dementia [22,45], although detailed investigation seems to indicate that slowed performance may be a task-specific finding [22]. Perceptual and psychomotor speed are normal throughout most of the clinical course of cortical dementia [16]. Examples of neuropsychological tasks for measuring aspects of perceptual and psychomotor speed are Digit Symbol of the WAIS(-R), Trailmaking test part A, Stroop Colour Word Test part I and II, symbol digit modalities tests and cancellation tests [57,90]. In recent studies use is also made of computerized reaction time tasks [22].

Attentional activities

In general attention, concentration, and vigilance are impaired in both cortical and subcortical dementia [16,17,22,26,66]. Auditory attentional span is intact, except in patients with progressive supranuclear palsy and in a lesser degree in patients with Huntington's disease in whom this (digit span forward) is also reduced [22,72]. Tests for measuring aspects of attentional activities include: Digit Span Test, Corsi Block Test, Mental control subtest of the Wechsler Memory Scale, subtracting serial sevens, part three of the Stroop Colour Word Test, Trailmaking Test, alternating sequences tests, continuous performance test, and cancellation tests [57,90].

Executive control functions

The term 'executive control functions' denotes the following cognitive functions: complex sequencing, mental flexibility, visual and verbal concept formation, verbal reasoning and judgement, visual sequential problem solving, and 'frontal lobe function' [22]. These functions can be examined with the use of neuropsychological tests such as: (card) sorting tests, the category test, Stroop interference task, Trailmaking test B, alternating cancellation tests, proverb interpretation, Comprehension and Similarities of the WAIS(-R), Raven's progressive matrices, the tower of London and comparable tasks, and parts of Luria's neuropsychological examination techniques [7,57,90]. Patients with Alzheimer's disease and patients with Pick's disease will show defective performance on these tests [14,66,83]. Patients with subcortical dementia also show impaired performance on these tasks [22]. Detailed analysis of performance on sequential subparts of tests or a qualitative analysis will reveal differences between cortical and subcortical dementia, and between subgroups of these syndromes [7,17,45,65,72].

Language

Language disturbances are often present in cortical dementia [14,16,26]. Empty speech and reduced word fluency is found in early Alzheimer's patients, and in early Pick's disease language impairment is relatively minor with semantic anomia and circumlocution [14,87]; in later stages of the disease Alzheimer patients develop characteristics of a fluent aphasia, while the use of language of patients with Pick's disease is reminiscent of frontal lobe impairment [14,87]. Also the syndrome of progressive aphasia can progress to a clinical picture of cortical dementia [36,37]. In patients with subcortical dementia a mild language disorder can be present, especially with regard to difficulties in performing more complex comprehension tasks, and reduced word fluency [22]; with regard to the latter, abandoning the time limit on these tasks can lead to improvement in patients with subcortical dementia, but not in patients with Alzheimer's disease [26].

Various neuropsychological tests are available for examining language functions such as general communicative ability, naming, word fluency, expressive and receptive language abilities, and repetition. These include the Boston Diagnostic Aphasia Examination, Communication Abilities in Daily Living, Porch Index of Communicative Ability, Token test [57,88,90], and the SAN - a Dutch Aphasia test [20,21]. In general, although handwriting can be impaired, no signs of agraphia are present in subcortical dementia. Handwriting is normal in patients with Alzheimer's disease, but features of agraphia can be present, even in an early stage of the disease [26]. Reading comprehension is normal in patients with subcortical dementia [22], but impairment can be present in patients with cortical dementia [14,15].

Contrary to patients with cortical dementia, patients with subcortical dementia have speech difficulties [14,22,26].

Arithmetic

Disturbances of calculation are evident in patients with Alzheimer' disease and patients with Pick's disease, and they may be due to primarily acalculia, or to visuospatial or aphasic impairments 14,26. Impairment of calculation in patients with subcortical dementia generally is less severe, and only becomes manifest on complex arithmetical tasks with time limits, or tasks requiring several mental steps in performing the calculation 22,26. Neuropsychological tests which can be used are: questions developed by Luria to test various aspects of arithmetical ability, Arithmetic subtest of the WAIS(-R)57,90, and the calculation subtest of the Groninger Intelligence Test 58.

Praxis

With exception of so-called gait apraxia in patients with normal pressure hydrocephalus, patients with subcortical dementia do not show signs of apraxia 22,45. Apraxia is present in patients with Alzheimer's disease and to a lesser extent in patients with Pick's disease 14,83. A useful scheme for examining different aspects of praxis is given by Poeck (1989)73.

Visual information processing

Visual information processing covers a wide range of cognitive abilities including visual recognition (with or without distracting elements), visual organization, and visuospatial ability. Examples of suitable neuropsychological tests are: incomplete figure recognition tests, Hooper Visual Organization test, Line orientation test, clock reading, Facial Recognition test, Unusual View, Raven Matrices 57,90. Visuoperceptual and visuospatial abilities are often evident in early stages of Alzheimer's disease 14,71,83. In Pick's disease only in later stages visuospatial ability will become affected 14,53. Basic visual gnosis is normal in patients with subcortical dementia. Minor impairment on visuoperceptual tasks and mild to moderately impaired performance on visuospatial tasks has been found in these patients 7,14,17,22, but usually only on more complex tasks which also make demands on the capacity for sustained attention and/or ideational-associative capacity 22. For differentiation between cortical and subcortical dementia one has to pay attention to the qualitative analysis of the performance on visuoperceptual and visuospatial tasks [26].

Visuoconstruction
Visuoconstructional performance combines perception with motor activity, always has a spatial component, and implies organizing activity [4,57]. Tests used for assessing visuoconstructional ability are: Block design of the WAIS(-R), three-dimensional block construction, Rey-Osterrieth Complex figure test, Bender Gestalt test, copying line-drawings and drawing to verbal command [4,57,90]. Patients with subcortical dementia can show very mild impairment on drawing tasks, and more prominent impairment on block-design tests. Defective planning or structuring seems the most probable cause of defective performance in subcortical dementia; contrary to patients with cortical dementia, patients with subcortical dementia can improve their performance when tasks are offered in parts [26]. Visuoconstructional skills are impaired early in the course of Alzheimer's disease and the performance of these patients reflects visuospatial impairment [14,66,71].

Motor speed
Slow performance on tasks measuring aspects of motor speed, such as finger tapping and peg board tests [57] is present in patients with subcortical dementia, and this can be a confounding factor in interpreting results on tasks requiring a motor response [22]. Bradykinesia is not one of the characteristic features of cortical dementia [14,16].

Mood
Personality and social behaviour remain remarkably intact during the early phases of Alzheimer's disease [14,16], although subtle changes may occur, consisting of accentuation of prior personality traits, apathetic attitude and distractibility [47]. In Pick's disease personality deterioration occurs in the initial stages, often including socially inappropriate activity and other behavioural disinhibition [14,87]. Changes described in patients with subcortical dementia include depression, apathy, inertia, irritability [14,17,83]. Depressive symptoms can be present in early stages of Alzheimer's

disease, but are less severe compared with those of subcortical syndromes [45]. In the examination of patients with dementia, apathy must be distinguished from depression, and the experience of emotion must be distinguished from its expression; the latter is especially important in patients with subcortical dementia, as they can be limited in their expression of emotions, due to impairment of facial and other movements. Well known instruments for determining the mood or affect during the course of clinical examination are questionnaires such as the Minnesota Multiphasic Personality Inventory and shortened versions of it, the Beck Depression Inventory, Zung Self-rating Depression Scale [10,44,57]. The difficulty with these instruments is that they rely on the ability of the patient to read and understand the test items, which can be difficult for patients with cortical dementia. An alternative is the use of rating scales, of which the Hamilton Rating Scale for Depression [39,40] is a primary example. One should, however, keep in mind that this rating scale was developed to measure the severity of depression and not to diagnose it [10]. Instruments such as the Cambridge Examination for Mental Disorders in the Elderly and the Dutch version of it (CAMDEX(-N))[24,76,77], the Geriatric Mental State Examination (GMS)[11,38,43], and the Diagnostic Interview Scale (DIS)[31,75] include items for diagnosing depressive and other psychiatric symptoms. Ratings of depressive symptoms in patients with neurological diseases should in general be interpreted with caution, because of a possible overlap of depressive and somatic symptoms [45].

Discussion and conclusion

Clinical neuropsychological research in dementia has primarily focused on the study and the assessment of cognitive changes, as their presence is important for diagnosis. An often mentioned criticism directed at studies examining the pattern of neuropsychological differences between various dementia syndromes concerns the assessment of the severity of dementia [45]. In many studies severity of dementia is estimated by the performance on cognitive (screening) tests, or is inferred from the score on a dementia rating scale. But it has become clear that not only cognitive impairment but also functional disability in daily life and behavioural disturbances must be important characteristics in the pragmatic assessment of the severity of dementia [85,86]. Hopefully in the future these measurements will be incorporated in the evaluation of patients with dementia and in objective studies examining the pattern of neuropsychological deficits between the various dementia syndromes.

To distinguish between various dementia syndromes, neuropsychological evaluation of cognitive function should be directed at assessment of intelligence, memory, perceptuomotor speed, attention, executive control function, language and language related functions, arithmetic, praxis, visuoperception, visuoconstruction, and motor speed. Also the presence and extent of personality changes and mood disorder should be evaluated.

Many dementia patients are taken care of by their families. The changes in cognitive capacities, personality and functioning in daily life of patients with brain damage also affect the well-being of those who have to take care of them [55]; this is also the case in patients with dementia [60,74,85,86]. To get a complete picture of the overall consequences of dementia attention should also be paid to the burden experienced by the caregiver [85].

The last 10-15 years there have been qualitative changes in the area and in the content of neuropsychology. Human neuropsychology is now in an era which promises to generate a deeper and more comprehensive knowledge of how the brain functions to mediate cognitive and affective behaviour. The focus of attention has shifted from 'topography' to concern with underlying neural and cognitive mechanisms [3,56,80,81]. An understanding of the relationship between brain and behaviour requires more than a correlation between behavioural dys-functions and lesion sites. The incorporation of the concept of modularity, currently dominating cognitive psychology, can provide a detailed functional analysis of cognitive deficits [81]. But to obtain a complete picture of the brain-behaviour relationship one not only has to know the impaired functional components of behavioural deficits, but one also has to search for the relation between these and the localization and extent of brain damage, and the functioning of the cerebral structures involved. For this

purpose, the discipline of neuropsychology must maintain a
readiness to relate, but not to subordinate, its behavioural
observations and theoretical interpretations to neuroradiological,
neurophysiological, neuroanatomical, and neurochemical data
6,70.

References

1. Albert MS.
    Assessment of cognitive dysfunction.
    In: Albert MS, Moss MB (Eds). Geriatric neuropsychology.
    New York, The Guilford Press, 1988: 57-81

2. Albert MS.
    Geriatric neuropsychology.
    J Consult Clin Psychol 1983; 49: 835-50

3. Benton A.
    Neuropsychology: past, present and future.
    In: Boller F, Grafman J (Eds). Handbook of Neuropsychology, vol 1.
    Amsterdam, Elseviers Science Publ BV, 1988: 3-27

4. Benton A.
    Visuoperceptual, visuospatial, and visuoconstructive disorders.
    In: Heilman KM, Valenstein E (Eds). Clinical Neuropsychology, 2nd ed.
    New York, Oxford University Press, 1985: 151-85

5. Benton AL, Sivan AB.
    Problems and conceptual issues in neuropsychological research in aging
    and dementia.
    J Clin Neuropsychol 1984; 1: 57-63

6. Broks P, Preston GC.
    Neuropsychological approaches to cognitive disorders - a discussion.
    In: Stahl SM, Iversen SD, Goodman EC (Eds). Cognitive
    Neurochemistry.
    Oxford, Oxford University Press, 1987: 191-200

7. Brown RG, Marsden CD.'Subcortical dementia': the neuropsycho-
    logical evidence.
    Neuroscience 1988; 25: 363-87

8. Butters B, Salmon DP, Heindel W, Granholm E.
   Episodic, semantic, and procedural memory: some comparisons of
   Alzheimer and Huntington disease patients.
   In: Terry RD (Ed). Ageing and the brain.
   New York, Raven Press, 1988: 63-87

9. Caine ED, Bamford KA, Schiffer RB et al.
   A controlled neuropsychological comparison of Huntington's disease
   and multiple sclerosis.
   Arch Neurol 1986; 43: 249-54

10. Christison C, Blazer D.
    Clinical assessment of psychiatric symptoms.
    In: Albert MS, Moss MD (Eds). Geriatric neuropsychology.
    New York, The Guilford Press, 1988: 82-99

11. Copeland JRM, Kelleher MJ, Kellett JM et al.
    A semi-structured clinical interview for the assessment of diagnosis and
    mental state in the elderly: the Geriatric Mental State Schedule. I.
    Development and reliability.
    Psychol Med 1976; 6: 439-49

12. Crevel H van, Teunisse S, Otten JMMB.
    Reversibele dementieën: meten van behandelingseffecten. (Reversible
    dementias: measuring the effect of treatment.)
    Tijdschr Gerontol Geriatr 1989; 20: 249-50

13. Crevel H van.
    Clinical approach to dementia. In: Swaab DF, Fliers E, Mirmiran M, et
    al (Eds). Aging of the brain and senile dementia. Progress in brain
    research 70.
    Amsterdam, Elseviers Science Publishers BV., 1986; 70: 3-13

14. Cummings JL, Benson DF.
    Dementia: a clinical approach, sec. ed.
    Boston, Butterworth, 1992

15. Cummings JL, Darkins A, Mendez M, et al.
    Alzheimer's disease and Parkinson's disease: comparison of speech and
    language alterations.
    Neurology 1988; 38: 680-84

16. Cummings JL, Benson DF.
    Dementia of the Alzheimer type: an inventory of diagnostic clinical features.
    J Am Geriatr Soc 1986; 34: 12-19

17. Cummings JL.
    Subcortical dementia: neuropsychology, neuropsychiatry, and pathophysiology.
    Br J Psychiatr 1986; 149: 682-97

18. Cummings JL, Benson DF.
    Subcortical dementia: review of an emerging concept.
    Arch Neurol 1984; 41: 874-879

19. Deelman BG.
    Neuropsychologische diagnostiek. (Neuropsychological diagnostics.)
    In: Luteyn F, Deelman BG, Emmelkamp PMG (Eds). Diagnostiek in de klinische psychologie. (Diagnostics in clinical psychology.)
    Houten, Bohn Stafleu Van Loghum, 1990: 136-61

20. Deelman BG, Liebrand WBG, Koning-Haanstra M, Burg van den W.
    SAN-afasietest: konstruktie en normering (herziene uitgave). (SAN-aphasia test: construction and norms.)
    Lisse, Swets & Zeitlinger, 1987

21. Deelman BG, Liebrand WBG, Koning-Haanstra M, Van den Burg W.
    SAN Test: een afasietest voor auditief taalbegrip en
    mondeling taalgebruik. (SAN-test: an aphasia-test for auditory language comprehension and oral language.)
    Lisse, Swets & Zeitlinger, 1981

22. Derix MMA.
    Dementia in diseases predominantly affecting subcortical structures: a neuropsychological contribution to the concept of subcortical dementia.
    This book, chapter 3.

23. Derix MMA.
    Illustrative case histories.
    This book, chapter 4.

24. Derix MMA, Hofstede AB, Teunisse S, et al.
CAMDEX-N: de Nederlandse versie van de Cambridge Examination for Mental Disorders of the Elderly (CAMDEX) met geautomatiseerde dataverwerking. (CAMDEX-N: the Dutch version of the Cambridge Examination for Mental Disorders of the Elderly with computerized data processing.)
Tijdschr Geriatr Gerontol 1991; 22: 143-50

25. Derix MMA, de Gans J, Stam J, Portegies P.
Mental changes in patients with AIDS.
Clin Neurol Neurosurg, 1990; 92: 215-22

26. Derix MMA, Groet E.
Subcorticale en cortical dementie: een neuropsychologisch onderscheid. (Subcortical and cortical dementia: a neuropsychological distinction.)
In: Schroots JJF, Bouma A, Braam GPA, et al (Eds). Gezond zijn is ouder worden.
Assen, Van Gorcum, 1989: 179-88

27. Derix MMA, Teunisse S.
Neuropsychologisch onderzoek bij ouderen. (Neuropsychological assessment of the elderly.)
Bijblijven, 1989; 8: 16-23

28. Derix MMA, Hijdra A.
Corticale en subcorticale dementie: een zinvol onderscheid? (Cortical and subcortical dementia: a useful distinction?)
N Tijdschr Geneeskd 1987; 131: 1070-4

29. Derix MMA, Hijdra A, Verbeeten BWJ.
Mental changes in subcortical arteriosclerotic encephalopathy.
Clin Neurol Neurosurg, 1987; 89: 71-78

30. Diesfeldt HFA.
Recognition memory for words and faces in primary degenerative dementia of the Alzheimer type and normal old age.
J Clin Exp Neuropsychol 1990; 12: 931-45

31. Dingemans P, Engeland van H, Dijkhuis JH, Bleeker J.
De 'Diagnostic Interview Schedule ' (DIS). (The Diagnostic Interview Schedule.)
Tijdschr Psychiatr 1985; 27: 341-59

32. DSM-III-R. Diagnostic and Statistical Manual of Mental Disorders, 3rd ed., revised.
    Washington D.C., American Psychiatric Association, 1987

33. Dubois B, Pillon B, Legault F, et al.
    Slowing of cognitive processing in progressive supranuclear palsy.
    Arch Neurol 1988; 45: 1194-9

34. Gallagher D, Thompson LW, Levy SM.
    Clinical psychological assessment of older adults.
    In: Poon LW (Ed). Aging in the 1980s: psychological issues.
    Washington, American Psychological Association Inc., 1980: 19-40

35. Gilson F, Hoogweg SJM, Andereoli PJH.
    Dat weet ik niet meer. Diagnostiek van dementie. (I don't remember. Diagnostics of dementia.)
    Almere, Versluys, 1990

36. Graff-Radford NR, Damasio AR, Hyman BT, et al.
    Progressive aphasia in a patient with Pick's disease: a neuropsychological, radiologic, and anatomic study.
    Neurology 1990; 40: 620-626

37. Green J, Morris JC, Sandson J, et al.
    Progressive aphasia: a precursor of global dementia?
    Neurology 1990; 40: 423-29

38. Gurland BJ, Fleiss JL, Goldberg K, et al.
    A semi-structured clinical interview for the assessment of diagnosis and mental state in the elderly: the Geriatric Mental State Schedule. II. A factor analysis.
    Psychol Med 1976; 6: 451-59

39. Hamilton M.
    A rating scale for depression.
    J Neurol Neurosurg Psychiatr 1960: 23: 56-62

40. Hamilton M.
    Development of a rating scale for primary depressive illness.
    Br J Social and Clinical Psychology 1967; 6: 278-96

41. Heindel WC, Salmon DP, Schults CW, et al.
Neuropsychological evidence for multiple implicit memory systems: a comparison of Alzheimer's, Huntington's, and Parkinson's disease patients.
J Neurosci, 1989; 9(2): 582-7

42. Hoch C.C, Reynolds CF.
Psychiatric symptoms in dementia: interaction of affect and cognition.
In: Boller F, Grafman J (Eds). Handbook of neuropsychology, vol. 4
Amsterdam, Elseviers Science Publishers BV, 1990: 325-334

43. Hooijer C, Tilburg van W.
Geriatric Mental State Schedule, GMS. Een psychiatrisch instrument in de psychogeriatrie. (GMS: a psychiatric instrument in psychogeriatrics.)
Tijdsch Geront Geriatr 1988; 19: 103-11

44. Hovaguimian T.
Instruments used in the assessment of depression in psychogeriatric patients.
In: Sartorius N, Ban TA (Eds). Assessment of depression.
Berlin, Springer-Verlag, 1986: 343-55

45. Huber SJ, Shuttleworth EC.
In: Cummings JL (Ed). Subcortical dementia.
New York, Oxford University Press, 1990: 71-86

46. Jenicke M.
Depression and other psychiatric disorders.
In: Albert MS, Moss MB (Eds). Geriatric neuropsychology.
New York, The Guilford Press, 1988: 115-144

47. Joynt RJ, Shoulson I.
Dementia.
In: Heilman KM, Valenstein E (Eds). Clinical Neuropsychology, 2nd. ed.
New York, Oxford University Press, 1985: 453-79

48. Kapur N.
Memory disorders in clinical practice.
London, Butterworth, 1988

49. Kasniak AW.
Neuropsychological consultation to geriatricians: issues in the assessment of memory complaints.
The Clinical Neuropsychologist 1987; 1: 35-46

50. King DA, Caine ED.
Depression.
In: Cummings JL (Ed). Subcortical dementia.
Oxford, Oxford University Press, 1990: 218-230

51. Kirshner HS, Tanridag O, Thurman L, Whetsell WO.
Progressive aphasia without dementia: two cases with focal spongiform degeneration.
Ann Neurol 1987; 22: 527-32

52. Knopman DS, Mastri AR, Frey WH, et al.Dementia lacking distinctive histologic features: a common non-Alzheimer degenerative dementia.
Neurology 1990; 40: 252-256

53. Knopman DS, Christensen KJ, Schut LJ, et al.
The spectrum of imaging and neuropsychological findings in Pick' disease.
Neurology 1989; 39: 362-368

54. Kolb B, Whishaw IQ.
Fundamentals of human neuropsychology, 3rd ed.
New York, WH Freeman and Company, 1990

55. Lezak MD.
Brain damage is a family affair.
J Clin Exp Neuropsychol 1988; 10: 111-23

56. Lezak MD.
Neuropsychological tests and assessment techniques.
In: Boller F, Grafman J (Eds). Handbook of neuropsychology, vol 1.
Amsterdam, Elseviers Science Publishers BV, 1988: 47-68

57. Lezak MD.
Neuropsychological assessment, 2nd. ed.
New York, Oxford University Press, 1983 (3rd. ed. 1994, in press)

58. Luteyn F, Van der Ploeg FAE.
Groninger Intelligentie Test. (Groninger Intelligence Test.)
Lisse, Swets & Zeitlinger, 1983

59. Luteyn F.
Intelligentietests. (Intelligence tests.)In: Luteyn F, Deelman BG, Emmelkamp PMG (Eds). Diagnostiek in de klinische psychologie (Diagnostics in clinical psychology.)
Houten, Bohn Stafleu Van Loghum BV, 1990: 124-35

60. Mace NL, Pabins PV, Castleton BA, et al.
The 36-hour day: caring at home for confused elderly people.
London, Hodder & Stoughton, 1985

61. Mandell AM, Alexander MP, Carpenter S.
Creutzfeldt-Jakob disease presenting as isolated aphasia.
Neurology 1989; 39: 55-58

62. Martin A, Brouwers P, Lalonde F, et al.
Toward a behavioural typology of Alzheimer's patients.
J Clin Exp Neuropsychol 1986; 8: 594-610

63. Massman PJ, Delis DC, Butters N.
Are all subcortical dementias alike? : Verbal learning and memory in Parkinson's and Huntington's disease patients.
J Clin Exp Neuropsychol 1990; 12: 729-744

64. Mayeux R, Stern Y, Spanton S.
Heterogeneity in dementia of the Alzheimer type: evidence of subgroups.
Neurology 1985; 35: 453-61

65. Morris F, Oscar-Berman M.
Comparative neuropsychology of cortical and subcortical dementia.
Canadian J of Neurol Sci, 1986; 13: 410-14

66. Moss MB, Alhert MS.
Alzheimer's disease and other dementing disorders.
In: Albert MS, Moss MD (Eds). Geriatric neuropsychology.
New York, The Guilford Press, 1988:145-78

67. Neary D, Snowden JS, Northen B, Goulding P.
Dementia of frontal lobe type.
J Neurol Neurosurg Psychiatr 1988; 51: 353-61

68. Neary D, Snowden JS, Bowen DM, et al.
    Neuropsychological syndromes in presenile dementia due to cerebral atrophy.
    J Neurol Neurosurg Psychiatr 1986; 49: 163-74

69. Nelson HE, O'Connel A.
    Dementia: The estimation of premorbid intelligence using the new adult reading test.
    Cortex 1978; 14: 234-244

70. Newcombe F.
    Neuropsychology qua interface.
    J Clin Exp Neuropsychol 1985; 7: 663-81

71. Orsini DL, Van Gorp WG, Boone KB.
    The neuropsychology casebook.
    New York, Springer Verlag, 1988

72. Pillon B, Dubois B, L'Hermitte F, Agid Y.
    Heterogeneity of intellectual impairment in progressive supranuclear palsy, Parkinson's disease and Alzheimer's disease.
    Neurol 1986; 36: 1179-85

73. Poeck K.
    Apraxie.
    In: Poeck K (Ed). Klinische Neuropsychologie, 2. Auflage
    Stuttgart, Georg Thieme Verlag, 1982: 188-207

74. Rabins PV, Mace NL, Lucas MJ.
    The impact of dementia on the family.
    JAMA 1982; 248: 33-335

75. Robins LN, Helzer JE, Croughan J, Ratcliff KS.
    National Institute of Mental Health Diagnostic Interview Schedule.
    Arch Gen Psychiatr 1981; 38: 381-89

76. Roth M, Huppert FA, Tym E, Mountjoy CQ.
    CAMDEX: The Cambridge examination for mental disorders of the elderly.
    Cambridge, Cambridge University Press, 1988

77. Roth M, Tyme E, Mountjoy CQ, et al.
    CAMDEX: a standardised instrument for the diagnosis of mental disorder in the elderly with special reference to the early detection of dementia.
    Br J Psychiatr 1986; 149: 698-709

78. Saint-Cyr JA, Taylor AE, Lang AE.
    Procedural learning and neostriatal dysfunction in man.
    Brain 1988; 111: 941-59

79. Schmand B, Louman J, Schaap T, Hoeks J.
    A quick low-budget method of test construction: The Dutch Adult Reading Test.
    J Clin Exp Neuropsychol; 1989: 11: 365
    (Schmand B, Lindeboom J, van Harskamp F. Nederlandse Leestest voor volwassenen (Dutch version of the NART). Lisse, Swets & Zeitlinger, 1992)

80. Sergent J.
    Some theoretical and methodological issues in neuropsychological research.
    In: Boller F, Grafman J (Eds). Handbook of neuropsychology, vol 1.
    Amsterdam, Elsevier Science Publishers BV, 1988: 69-81

81. Shallice T.
    From neuropsychology to mental structure.
    Cambridge, Cambridge University Press, 1988

82. Spreen O, Tuokko AT.
    The neuropsychological assessment of normal and disordered cognition.
    In: Malathesa RN, Hartlage LC (Eds). Neuropsychology and cognition, vol.1.
    The Hague, Martinus Nijhoff Publ, 1982: 63-112

83. Strub RL, Black FW.
    Neurobehavioral disorders: a clinical approach.
    Philadelphia, FA Davis Company, 1988

84. Swieten van JC, Geyskes GG, Derix MMA, et al.
    Hypertension in the elderly is associated with white matter lesions and with cognitive decline.
    Annal Neurol 1991; 30: 825-30

85. Teunisse S, Derix MMA, van Crevel H.
Assessing the severity of dementia: patient and caregiver.
Arch Neurol, 1991; 48: 274-77

86. Teunisse S, Derix MMA, van Crevel H.
Het meten van de ernst van dementie: patiënt en verzorgers.
(Measuring the severity of dementia: patient and caregivers.)
Ned Tijdschr Geneeskd 1990; 134: 327-31

87. Tissot R, Constantinides J, Richard J.
Pick's disease.
In: Frederiks JAM (Ed). Handbook of clinical neurology, vol 2.
Amsterdam, Elsevier Science Publ, 1985: 233-245

88. Van Dongen HR, Van Harskamp P, Luteyn F.
Tokentest: handleiding. (Tokentest: manual.)
Nijmegen, Berkhout Nijmegen b.v., 1982

89. Warrington EK.
Recognition memory Test: manual.
Windsor, NFER-Nelson, 1984

90. Weintraub S, Mesulam M-M.
Mental state assessment of young and elderly adults in behavioral neurology.
In: Mesulam M-M (Ed). Principles of behavioral neurology.
Philadelphia, F.A. Davis Company 1985: 71-123

91. Wilson RS, Rosenbaum G, Brown G.
The problem of premorbid intelligence in neuropsychological assessment.
J Clin Neuropsychol 1979; 1: 49-53

Index